Truth in the Sciences

Truth in the Sciences

SCOTT BUCHANAN

UNIVERSITY PRESS OF VIRGINIA
CHARLOTTESVILLE

THE UNIVERSITY PRESS OF VIRGINIA

First published 1972

The title page illustration is from
The Elements of Geometrie of the Most Auncient Philosopher Euclide of Megara.
This is the first edition in English, published
in London in 1570. The illustration is used with the courtesy
of Bennett & Marshall, Los Angeles, California.

ISBN: 0-8139-0388-2
Library of Congress Catalog Card Number: 78-176077
Printed in the United States of America

Contents

Foreword

SCOTT BUCHANAN'S *Truth in the Sciences,* written in 1950, is here published for the first time. It was to have been one of a set of essays by different authors, to be published together as a kind of guide or *vade mecum* for the use of the common reader of the classics of the western tradition—an addressee to whom Buchanan, with democratic leanings and in pointed revolt against some of the vanity and pedantry and narrowness of academe, had directed one large measure of his life's thought and concern. The joint project failed of actualization, and Buchanan's essay was still a manuscript in a drawer at the time of his death in March, 1968. On examination it proves to have very little of the Baedeker about it, and very much of what those who knew him will recognize as essential Buchanan. It is a meditation on the sciences in their relation to the past and the destiny of the human speculative enterprise as a whole; and it is an after-image of many dialogues pursued in the faith that, in Buchanan's words, "the reason that seems to guide history in a dream is neither asleep nor fully awake, and it will wake itself only by trying to talk." Buchanan saw modern civilization as having lost its vital unity in noncommunicating, strait and narrow ways and as having produced corresponding splits in its individual members; and the remedy, as he inferred from the diagnosis, could only be a dialectic that did justice to all the parts that had become separated, by showing their relation to the whole from which they derived. Of this therapeutic concern the essay before us was one outgrowth and expression.

During his lifetime Buchanan often heard his ideas branded as dilettantish and amateurish, or as dangerous and corrupting to youth, capable of turning them into misfits. Amateur and dilettante he gladly acknowledged himself to be, accepting these terms in their etymological meanings: he was a "lover" of ideas and he "delighted" in them, following in conversation their interconnections with loving delight. That the educational process as he engaged in it and guided it could turn young people into misfits, at least temporarily, he also acknowledged, but he urged that such misfitting was a necessary stage if the rifts in society and culture and individual souls were to be observed from the right distance for understanding. He believed revolution was necessary, continuous, and unavoidable, and he spent his last years thinking about how to legitimize it.

What was the essential Buchanan? "A teacher of the Socratic persuasion" is the description he gives of himself at the start of the essay. He had in fact a special capacity for "I-and-thou" encounter, as his former students know; the arts of listening and of questioning had become part of his fiber, and with irony and playfulness he blended a quality of attentiveness that made students feel he understood them. This way of teaching and learning had come partly by nature and partly by learning. Early on he had made something like the Socratic return to the logos. It was Alexander Meiklejohn, president of Amherst during his student days there, whom he blamed for setting him on the Socratic path of dialogue; the Platonic writings helped fix the direction, and in the company of friends, students, and enemies who sometimes became friends, he pursued it through a lifetime.

But in the modern world, with its imposing achievement of scientific knowledge and know-how, the Socratic course is not obviously open nor obviously the serious possibility that Socrates had taken it to be, and the right to pursue it may need to be rewon. For Buchanan one source of liberating insight seems to have been Kant, whose writings he began studying while a Rhodes scholar at Oxford. Kant, having acknowledged the legitimacy of the causal explanations of science in every realm of phenomena, including the phenomena of human behavior, had gone on to argue that man was nevertheless in principle the legislator of his own acts. Since Kant's day the world had turned post-Newtonian with Einstein's revolution, and the strict Kantian orthodoxy that held such a revolution to be impossible had itself to be rejected; but on the other hand, a simple extension of the Kantian direction of thought could accommodate revolution by suggesting that man was legislator of his theoretical as well as his practical acts. Buchanan's first book, originally a doctoral dissertation written for the Department of Philosophy at Harvard but then expanded and in 1927 published under the title *Possibility,* was such an extension; he here took a long step backward from theoretical systems in general and began a search for a general methodology, "an organon of intellectual imagination," as he called it. Faced with the mutual opacities of modern systems and complexes of thought, he set himself the task of discerning and disentangling the common features of systems in general that tend to remain invisible—features often embedded in the ways of language and particularly in the roles of analogy in poetry and mathematics and ordinary speech.

This was one aspect of Buchanan's modern return to the logos, and it dovetailed with other outcomes that were more deeply Socratic and Kantian or post-Kantian as well. The long step backward led to a detached vision of system-making in general. In anyone's mind as well as over the centuries, systems expanded and pretended to world-hood; and every world, as Buchanan saw it, was something of a poem, something made, a hypothesis. This conclusion was skeptical, to be sure, but as in the Socratic discovery of ignorance, it was not the dark end of a blind alley: if worlds are hypotheses, Buchanan reasoned, then "they are only possibilities, which with boldness, laughter, and ingenuity on our part can be put aside and replaced." And this conclusion was followed by another, Kantian in immediate inspiration but still parallel to a Socratic counterpart: "If we allow our minds their speculative indulgence, even to the limit of a Satanic multiplication of such worlds, a heavy duty is laid upon the practical intellect to weigh these worlds, to choose the best of them . . . and to will it." For Buchanan as for Socrates, there was thus an articulation between the speculative and the moral. In neither of their cases was it simply a matter of logic, nor was it quite free of mystery, but in both it effectively determined the directions of inquiry and action.

During the years 1925–1929, Buchanan was assistant director of the People's Institute in New York, engaged in planning and directing adult education programs of lectures and seminars. A concern that grew in the context of the seminars of these years was to bridge from one direction or another the gap between the sciences and the humanities, the "two cultures" as Snow has more recently christened them. For the ordinary seminar participant as for many a professional philosopher, mathematics and the mathematized sciences with their pervasive roles in modern culture were operating to produce a kind of complex or center of anxiety: "we admire and mistrust," Buchanan noted, "what we do not understand." Because of mathematics, science had become the secret doctrine of a cult—a cult somewhat comparable, Buchanan thought, to the hated and feared cult of Christianity whose rise and spread Gibbon had seen as a cause of the decline and fall of the Roman Empire. Full citizenship in the modern world demanded not mere entry into the cult but naturalization of its foreignness.

In a search for accessible routes across the barrier, Buchanan turned back to explore the ancient and medieval tradition of the liberal arts, noting the shifting relations and subordinations of one art to another in different historical periods, and seeking a modern equivalent that would embody the insights of the earlier tradition while encompassing the new analytic arts that had come to the fore with the scientific revolution. During these same years Buchanan was beginning to supplement the list of great books studied in the adult seminars at the Institute—a list which had been taken over from a Columbia University honors course and was nearly innocent of mathematics and natural science, but which Buchanan wanted to see expanded so as to give the sciences a present voice in the conversation of which they were neglected or forgotten parts. Fully half the relevant background of Dante's *Divine Comedy,* he liked to argue, is in Euclid's geometry and Ptolemy's astronomy; and at the heart of Newton's *Principia,* he would insist, is a poetic vision of the world. Poems and novels may be penetrated by mathematical insight; and there is also a reverse illumination that comes from reading the scientific books as if they were poems, narratives, and histories, which, he would add, is what they are, so long as the human mind is discursive. Some of these views and insights found expression in published works: in *Poetry and Mathematics* (1929; reissued in 1962) and in *Symbolic Distance* (1932), Buchanan used the mathematical notions of matrix and projection in an analysis of metaphors and fictions in poetry and science; and in *The Doctrine of Signatures* (1938) he sought light on the unity of knowledge as it could be glimpsed in another perspective, that provided by the discipline of medicine.

These various but related lines of study continued to occupy Buchanan during his years as professor of philosophy at the University of Virginia (1929–1936). Then in 1936 he was called by Robert Hutchins to Chicago to organize and direct the Committee on the Liberal Arts. The task that Hutchins set the committee was to find what could be done amidst the chaos produced by the elective system to recover the liberal dimension of liberal education. The committee imagined a curriculum, and in the following year Buchanan, along with Stringfellow Barr, who had accompanied him from Virginia to Chicago, was challenged by the trustees of St. John's College in Annapolis, Maryland, to put the committee's imagined curriculum into practice. Buchanan became dean and Barr president of the college, and for the next nine years they labored at the creation of what came to be called "the New Program" or later simply "the Program." Those years, Buchanan acknowledged to the present writer a long time afterward, were arduous when not stormy, but he looked back upon them as his best. The aim for St. John's was no more and no less than to bring into being a community of learning in which the primary assumptions and roots and the mutual interconnections and relevances of the major classics of the western tradition were brought into a lively and continuing discussion.

However mildly described, this was revolution, and it involved difficulties and frustrations and battles as revolutions tend to do. There were questions of detailed and large-scale planning on the "domestic" side, and problems of "foreign relations" that bore on the survival of the Program and its role in what some, both inside and outside the college, dreamed could become an American renaissance. The dream, as everyone knows, did not materialize. The Program, however, managed to survive, weathering crises, developing and changing in certain ways, but moving generally in the direction Buchanan had set. Dismissed were electives, intercollegiate athletics, the survey course, and the watered-down textbook. The central feature of the curriculum became the twice-weekly seminar on a great book. This was supported and complemented by daily exercises in translation from foreign languages and in mathematics, and by twice-weekly sessions in the laboratory. In all these endeavors there were failures and successes, dependent on the choice of subject matters and on what the faculty and students could bring in the way of background and backbone. A major outcome was the emergence of a community singular in its kind, a college committed to the idea of liberal education in its traditional sense, insistent on the unity of this idea, and insistent on the communal character of the means to its realization.

What about the "two-culture" problem, the problem of bridging the gap between the world of the sciences and the world of the humanities? In Buchanan's view, one major part of this problem was the amnesia from which the sciences were suffering with regard to their own history and roots, and another was the climate determined by the universal use of gadgets and machines, the use of the word *scientific* in the rhetorical sense of magic, and the half-hidden technological fabric of our lives—a climate which left a man's mind fogged and clogged with the images and myths of science, shadows and reflections of things he could not grasp. The attack on the problem was waged from several sides. There was extensive exploration of works from the past in mathematics and natural science, to locate and reconnoiter the classical ones, those that like literary works had beginnings, middles, and ends, and moved from familiar situations through complications to unravellings and recognitions. There were also intensive efforts to set up meaningful sequences of laboratory exercises, the aim here being not primarily the mastery of scientific information but rather insight into the interpenetrations of the operations of hand and eye and mind in what is appropriately from the Latin called "fact," the thing done or made. Such questions were pursued as: How, and with what instruments, could you construct "from the ground up" the units and the laws of electromagnetism? How could you go about constructing a plane surface or a slide rule or a set of weights for the analytical balance if you found yourself stranded in the primeval forest and had to begin all over again?

Some particular undertakings and outcomes are worth mentioning. Under Buchanan's sponsorship, two of the tutors of the college undertook translations of classics that had been more frequently referred to than read: R. Catesby Taliaferro translated Ptolemy's *Almagest,* and Charles Glenn Wallis translated Copernicus's *On the Revolutions of the Heavenly Spheres* and parts of Kepler's *Epitome of Copernican Astronomy* and *Harmonies of the World.* All of these were "firsts" as English translations; scholars have found some faults in them, but three decades after their completion in the late thirties no new translations have supplanted them or yet been undertaken. They opened up to the student the possibility of seeing how the hypothetical circles of Ptolemaic astronomy are introduced in the analysis of the phenomena; how the Copernican revolution in its main outlines simply carries Ptolemaic astronomy through a geometric transformation; how the perspective relations and systematic unity introduced by this transformation inspired in Kepler a new vision of man and of the world and a new program of investigation. From this sequence one learned respect for the Ptolemaic hypotheses, seeing the degree of directness with which they represented the observational data; one learned some elementary planetary astronomy and saw the indispensability of hypotheses for any decision about planetary paths and speeds.

As starting point for the study of this and other parts of the mathematical and scientific tradition, Buchanan turned to Euclid's *Elements,* and this feature of the Program proved in many respects regularly and remarkably successful. Only about a third of the *Elements* was being studied in schools elsewhere, and then only in secondary versions. Buchanan saw this work as having an elegance of total structure and a clarity of detailed construction that made it a model of intellectual beauty—one that he liked to say had been sweetly and gently informing all European thought up to the middle of the last century. The fifth book of the *Elements,* long neglected in elementary instruction, was truly the high point of Greek mathematical sophistication, Eudoxus's theory of proportions, embodying an elegant and logically rigorous way of coping with the Pythagorean *scandale* of incommensurability. It now became a landmark and point of comparison in studies of the way in which Vieta and Descartes in the seventeenth century, with their attention focused on operations, bypassed the problem of incommensurability, and of the way in which Dedekind in the late nineteenth century, with formulations closely parallel to Euclid's, returned to the Greek standard of rigor. But more important for the student than such points of relevance to later developments was the luminous clarity of the work itself, the feel it gave for logical demonstration, the recognition it brought of an artful unity of arrangement culminating in the construction of the five regular solids. Euclid is supposed to have said there was no royal road to geometry, but for many of the students who came to St. John's his own book proved a master key, opening doors they had assumed closed to them.

The Program led on to more difficult works; two that produced much shaking of heads and continuing debate were Apollonius's *Conics,* of which Taliaferro translated the first three books, and Newton's *Principia.* Professional mathematicians and physicists asserted dogmatically that the first of these should not and the second could not be read, even by themselves, because—*nous avons changé tout cela.* To Buchanan it seemed shameful that we should live for the most part in a familiar Newtonian world and be unable to read the book from which that world emerged. Books that seemed unintelligible to both teacher and student, he insisted, could become approachable and conquerable if the proper path through other books were followed. In the case of Apollonius and Newton, paths across the terrain were at length found, and for those who persevered they led to discoveries and views that would not have been met with along the more usual textbook roads of today. In the artful intricacy of Apollonius's work were not only propositions upon which Newton's *Principia* depended, but also analogies and themes and problems that could now be seen as the lost background and palimpsest behind other seventeenth-century innovations—the analytic geometry of Descartes and the projective geometry of Desargues and Pascal. In the *Principia* one could follow the roles of concepts like "force" and "quantity of matter" and "universal gravitation" independently of their more recent sedimentation in familiar algebraic formulae, and one could view the universe that these concepts produced as meaningful construct rather than as brute fact. Farther along the path one could reach comprehensive views that ordinarily and for those who followed the now customary algebraic-style routes remained unglimpsed; it is an impressive outcome, for instance, to see the lunar anomalies, the precession of the equinoxes, and the general phenomena of the tides emerging in qualitative detail as cases of a single theorem. The Program was operating at its best when such particular results were taken not as truths simply to be remembered or applied to cases but as occasions for considering aspects of things not yet considered, meeting views of the world not encountered before, discovering the familiar in the new and the questionable in the familiar. Sometimes it happened.

What about the other books through which, in Buchanan's programmatic vision, a pathway was to be found leading to citizenship in the modern world? His original list had included not only Copernicus and Harvey and Galileo and Descartes; it had gone on to Fourier's *Mathematical Analysis of Heat,* Faraday's *Experimental Researches,* the *Mathematical Papers* of Gauss and Galois, Maxwell's *Electricity and Magnetism,* Hamilton's *Quaternions,* Cantor's *Transfinite Numbers,* Russell's *Principles of Mathematics,* Veblen and Young's *Projective Geometry,* and more besides. Whether with love or irony, one must admit: the list had heroic dimensions. It was, in fact, too long, too demanding, for the available time and patience and knowledge and guidance. Worse yet, even as it stood, it might not be long enough, might lack some of the links required to make a proper chain of learning. *That* question could be answered only through research by a team of scholars, and research in turn required time and cash, both of which were in short supply once the undergraduates had converged on the campus, and were waiting there in the classrooms to see what all this talk about liberal education amounted to. The result was a bit of "tragic recognition." The list had to be shortened. Then as the years of World War II came on, there was a further source of difficulty and confusion, the manpower drain and consequent turnover in faculty, which sometimes—and particularly in the case of the laboratory—had the result that very imaginative ideas that had once been tried and proven were lost in the shuffle. Some shortcuts, textbooks that were certainly not classics, were introduced with reluctance or in desperation.

The present writer joined the St. John's faculty in 1948, after the first period of innovation had passed, and the war years had been survived. Barr and Buchanan had left to try, abortively as it turned out, to establish a sister college in Massachusetts, and then had gone on to another project, the Foundation for World Government, whose major concern became the economic development of the unindustrialized countries. Meanwhile St. John's had entered what some called its "Alexandrian" period, to distinguish it from the first, "Athenian" period; there was a half-nostalgic admission that the Hellenic had passed over into the Hellenistic. What this signified, as far as a novice could make it out, was that the place was haunted by the memory of some departed Socratic spirit, and moreover, that there was general if reluctant acknowledgment that the routine, pedestrian sorts of learning, the grammars of the various subject matters studied, must be given more steady and conscientious attention from now on.

Actually, the Socratic daemon was still about, saying "No" to much mere ingestion and regurgitation of information that collegiate education too often got reduced to, and insisting, if anyone claimed knowledge, that he submit himself to cross-examination. During the next ten years I gave much of my effort to the laboratory instruction, as did a few other tutors. We were on a sort of firing line, an exposed front of the "two-culture" war. In a program of which each part was required for every student, and in which no part could escape the obligation of trying to justify and defend itself as liberal education, it was a persistent and much-argued question what the laboratory course should or could become, whether and how it could move beyond such standard instruction in physics, chemistry, and biology as was generally going on elsewhere to something that could justifiably be required of all students—something that it would be generally agreed had an identifiable liberal dimension, a recognizable and articulable relevance to thought about the processes of man, "the knower." If there were occasions when we had some measure of success, it was by asking Socratic questions, by remembering that even a "fact" is no simple thing but the product of a mind which actively comprehends what it sees, by proceeding to explore fact and theory in their interrelation. But even when we seemed to be proceeding in the right direction, we did not always manage to persuade the bright young liberal artist to descend from the dialectical epicycles in which he might be revolving rather splendidly, to examine, say, the mere material assemblage of condenser plates, motorized switch, and ballistic galvanometer from which, so beautifully as I thought, there emerged the constant in Maxwell's wave equation, the speed of light. At such moments I would be tempted to ask: what in the name of the nameless had Buchanan been after?

He liked to point out that Socrates' famous assertion, "I know that I do not know," as it was first read or written down without the diacritical marks introduced by modern Greek scholars, had to be interpretable as also meaning "I know *what* I do not know," or even "I know *whatever* I do not know." The punning paradox was the Socratic reply to the difficulty Meno had raised in the dialogue about virtue: "Even supposing, at the best, that you hit upon this thing you are seeking, how will you know it is the thing you did not know?" It was a hypothesis for a life of inquiry. And in the last analysis, Buchanan felt, the inquiry could not limit itself to this or that. In 1958, speaking to a gathering of former students and others who had participated in the St. John's undertaking during the Barr and Buchanan era, he said, or asked:

Under the slings and arrows of outrageous fortune, have you persuaded yourself that there are knowledges and truths beyond your grasp, things that you simply cannot learn? Have you allowed adverse evidence to pile up and force you to conclude that you are not mathematical, not linguistic, not poetic, not scientific, not philosophical? If you have allowed this to happen, you have arbitrarily imposed limits on your intellectual freedom and you have smothered the fires from which all other freedoms arise.

Most of us have done this and come short of what that threadbare slogan, human dignity, really means. We are willing, and shamefully relieved, to admit that each has his specialty, his so-called field, and the other fellow has his, and we are ready to let the common human enterprise go by default.[1]

As to the "common human enterprise," Buchanan had some further words that threw light on what he meant by it:

In *The Brothers Karamazov,* Ivan tells Alyosha that he finds it easy to believe in God, but that he finds it impossible to believe in the world. For most of us these days, we have believed in some things so weakly or fanatically that other equally or more real things have become absurd or impossible. This results from our crippled minds, our self-imposed limits on understanding, our deafness to the choice that asks: Is it true? I am persuaded that the cure for this sickness of mind is in some vigorous and rigorous attempt to deal with that most puzzling and mysterious idea, the idea of the world. It is not a simple idea, nor even a merely complicated idea. Kant called it an antinomy, an idea of speculative reason governing all other uses of the intellect. There have been other such ideas that have governed thought, the idea of God or Being as it puzzled and dazzled the ancient world, the idea of Man as it stirred and fermented the world from the Renaissance on. God and Man have not disappeared as charts and aids to intellectual navigation, but they are in partial eclipse at present, and the world is asking us the big questions, questions in cosmology and science, questions in law and government. They are not merely speculative questions; they are concrete and immediately practical. They are as much matters of life and death and freedom as the old questions were. Most of us have made, with Ivan, a pact with the devil, an agreement not to face them and accept them—yet.[2]

Buchanan spent his last years at the Center for the Study of Democratic Institutions in Santa Barbara, thinking and writing about law, revolutions in general, and the present revolution of the world in particular. Revolution, he pointed out in his last essay, is a natural property of the world: "the world lives and moves, it revolves."[3] But for many years he had been viewing the revolution of the present age with special wonder, gauging its extent and formulating hypotheses about its import. Earlier than most he had been thinking and speaking about the significance for the world of the "underdeveloped" countries, the truly developing ones. Earlier than most he had been thinking and writing about what technology had done and would do to civilization. Earlier than the university students of today he had worried about what he called the "withering of consent," the fact that more and more

[1] *Embers of the World,* ed. Harris Wofford, Jr. (Santa Barbara, Calif.: Center for the Study of Democratic Institutions, 1970), p. 2.

[2] *Embers of the World,* pp. 2–3.

[3] "A Message to the Young," *The Center Magazine,* I (March, 1968), 8–13.

of the political and economic decisions in our society were being made by default or in accidental ways, without the debate and understanding and meaningful consent of the governed, and he had proposed some remedies in his *Essay in Politics* of 1953. His view of the present was mixed with no naïve optimism about particular remedies or about the future prospects of civilization; he kept a print of Dürer's *Melancholia* above his desk.

Yet he had a faith, too, and it was a strong one, in the intellect in each human being as it could be awakened to its proper functioning through the process of Socratic inquiry. In the very special time in which we live, in the sixties as well as the twenties and thirties of this century, he saw such awakening as a preeminently needful thing. It meant asking "Is it true?" at unexpected times. It meant seeing the necessary dispersive moments and stages of knowledge as steps in the common march to intelligible unity. It meant entertaining the idea of the world, and attempting to accept the responsibility that such entertaining implied. Buchanan from somewhere out on the rim of the world was always trying to point to the whole of things. It was with such an intention that *Truth in the Sciences* was written.

In preparing the manuscript for publication, I have made a few minor corrections in the text. All of the notes are my own; they are intended to explain some of Buchanan's allusions that might otherwise remain obscure, to give explicit references for the paraphrased doctrines of various authors, and on occasion to supply additional information about the topic under discussion. On a few matters, I have begged to differ with the author; his way of analyzing the Darwinian hypothesis is the main case in point. For assistance in a careful reading of the text and in preparing the editorial notes, I owe much to my students Philip Chandler and James Liljenwall.

July, 1971

Curtis Wilson
Professor of History
University of California, San Diego
formerly Tutor and Dean
St. John's College, Annapolis

Introduction

I AM not a professional scientist, although I would like to claim a solid amateur standing in terms of a working curiosity in the persistent problems of the scientific enterprise. This rightly suggests a philosophical interest, but I cannot claim unbroken devotion to that academic guild. Although I have the philosopher's anxiety about the partial successes, frustrations, and ultimate destiny of the human speculative enterprise as a whole, the focusing mirrors that I have used have been borrowed from various other disciplines of enquiry, first from mathematics and physics, then from medicine, theology, and law in that order over a period of thirty years. I am not aware of deliberate membership in any school of philosophy, but I am accused of having a Platonic bent of mind, an accusation which I accept as confirmation of my deep aversion to party philosophies. But I am by will, and now by confirmed habit, a teacher of the Socratic persuasion. I believe that all men have intellects that are awakened and freed for their proper theoretical and practical functions by the process of question and answer, that this process can be started by an external teacher and becomes self-perpetuating as an internal process in a society where conversation and deliberation are respected.

A seminar in the discussion of great books is such a society and has an inherent tendency to expand. The happiest hours I have spent have been in the discussion of the great books with college undergraduates, with graduate students, and with adults. It is out of a memory of these occasions that I have written this essay. I have recalled things that have been said and at times have tried to imagine what might be said by the extremely various persons who have done or will do the common reading. It would have been better had I been able to write in the dialogue form, but the difficulties have been great enough, if not too great, in this second-remove and somewhat nostalgic reporting of remembered and imagined conversations. It will be minimally successful communication if the reader finds beginning points for actual conversations which he will have. He should follow the conversation wherever it leads.

Although it may not be true of all scientists, it is true of the great scientist that after being a human being, he is first of all a poet. There was a time in his life when he had a poetic vision of the world. Without entering into any theories of inspiration and revelation to determine the sources of poetic visions, an examination of their form and content shows that they are made of the matter of memory formed and transformed by inventive imagination. Their remarkable properties for our purposes can be summarized by saying that they are worlds, that is, that they have a stout systematic internal structure and that they have an independent existence. They are therefore intelligible in themselves without any need for reference beyond themselves.

On the other hand, the poetic vision is obviously the "imitation of an action serious, complete, and of a certain magnitude,"[1] the limited work of a human maker with all the luminous and suggestive power of a myth. The Pygmalion who makes such an artifice finds it full of life and movement, and in wonder at his own miracle falls in love with it. He finally comes to think of himself as made by it and as the wandering, searching agent of its inspiration. His adventures, his curiosities, his observations, and his further inventions seem to flow from his artificial muse.

If this is true of the great scientist, as well as of many other men who would not style themselves poets, it would seem that there is a master key, if not a royal road, to the science which flows from and lives in the light of the vision. Not all the writers of the great books in science obviously provide us with the key; sometimes a philosopher or another scientist lets the secret out. There are five cases of this that I should like to point out to the reader for what they are worth.

The first is in Plato's *Timaeus,* (27c–34c) where on his own word he is discerning the vision that lay perhaps in the stars for the Pythagorean brotherhood. It is the description of the materials, forms, and conditions of work of the great artificer, the Demiurge. Its light still shines in the worlds of mathematics and physics.

The second is in Lucretius's *On the Nature of Things* (I, 950–1107), where he is acting as poetic midwife for Democritus and Epicurus, who first saw a world freed from artificers and gods, taking care of itself or carelessly floating free in the void.

The third is from Dante's *Divine Comedy* (*Purgatorio,* 76–139), where he epitomizes all of the Aristotelian and scholastic world in the lyrical formulation of the principle of natural love. Book Lambda of Aristotle's metaphysics may be better poetry and better philosophy, but Dante's Virgil is probably a better teacher for the beginner.

The fourth is from Gilbert's *On the Magnet* (Book V, Chapter XII) and illustrates that the vision may be poetic although the language

[1] Buchanan is here quoting part of Aristotle's definition of tragedy (*Poetics* VI. 1449b24–25).

may not be up to the standard of the three preceding selections. In this case the reader is persuaded that Gilbert had the vision long before he made his searching observations and experiments, and that the systematic work was passionate as well as thorough. The light from this vision floods the work of Kepler and seems to be a fire kindled near the newly recognized center of light, life, and motion in the sun. It has been lighting new sources of search in the centuries since.

The fifth is from Newton (*Principia,* General Scholium) who said that he had been merely playing with pebbles on the beach of a vast ocean, and that he was a pygmy standing on the shoulders of giants, but it should not be forgotten that he ended his life working on the chronology of the Bible. It should not be too surprising that he called space God's sensorium.

The literary critic will not think that all of these and many more that might be cited are great poetry. The reader may think that he is being invited to become a child. But these passages can do two things for the adult who may be willing to become childlike. They will gently lift him to the shoulders of some of the giants where he can see the beacons of the scientific enterprise, and they may start a habit in him of the kind of reading that every man must do in science if he is to become a full citizen of the modern world. Poetry is a medium of knowledge, and it can save the world both from superstition and from the barbarism of the learned cults.

There is another beginning point for the common reader, the great books in medicine. There are three great authors to be mentioned here, Hippocrates, Galen, and Claude Bernard, but their books should be taken only as probably the best samples of a great historic literature which is now being recovered, catalogued and edited for the use of the medical profession, which needs to become aware of its long and significant history. Much of this history is important for the lay person because it records almost continuously the application of science to men, who through medicine become the objects of both scientific study and scientific practice.

It would be too much to say that medicine is the fulfillment of the Delphic imperative, Know thyself, but it is through medicine that the whole scientific enterprise is reminded of its humanistic end. It is likely that medicine is the most frequent locus of renaissance, but it is more than likely that it was a prime mover in the rebirth of science during the Italian Renaissance. Galileo, Gilbert, and Harvey all had medical training, much of it in the medical school at Padua.[2] Pope's great

[2] Strictly speaking, of the three scientists Buchanan names only Harvey is known for sure to have studied medicine at Padua. He studied there under Fabricius of Aquapendente, whose discovery of the valves of the veins was

slogan in the *Essay on Man,* "the proper study of mankind is man," can be seen as a later reflection cast by the focusing of the Renaissance on medicine and on man.

This interpretation of the Delphic oracle may seem all too literal, but the contents of medical books and the concern of medical thought throughout the tradition are not as restricted as some overemphasized parts of it seem to indicate. In one form or other the most persistent article of faith in the tradition is that the universe, all that is, is a macrocosm mirroring man, the microcosm. It is the exploration of this cosmic reflection for the sake of both theory and practice that starts and corrects the scientific enterprise. In fact, it is the reflection of the big universe in the little universe, the mind of man, that constitutes science. It is therefore not surprising that the end and final test of science is its application to the individual man. Both his wisdom and his body are involved in such a way that medicine is the science of the soul and science is the medicine of the soul.

But there is a special application of these general considerations. Medicine may be the constitutional government of the sciences, but it stands outside them as they set up their separate shops, and it has a special task which they can ignore or at least postpone. The physician must use science to understand and aid the individual. A particular science may start with the concrete and return to it in the end, but it lives chiefly in the general and the abstract. The physician must use abstractions and generalizations, but he cannot stop in them and let them seem merely good in themselves; he must see them in his case. Therefore, the terms of medical knowledge are not observation, prediction, and verification; they are diagnosis, prognosis, and therapy. His place of knowledge and practice is the clinic or the bedside, not the laboratory, the observatory, or the study. On the other hand, the clinic

---

one of the important steps leading to Harvey's discovery of the circulation of the blood. In the case of Galileo, who taught at Padua and whose physician there was Fabricius, the influence of medical studies is more debatable. As a young man he seems to have disrelished what he was learning at the University of Pisa in preparation for a medical career, and at age twenty-one abandoned medicine for mathematics. However, an argument connecting the 'resolutive-compositive' method of Galileo's *Two New Sciences* with discussions of scientific method in the medical schools of northern Italy, and particularly in the School of Padua, has been given by John Herman Randall, Jr., in "The Development of Scientific Method in the School of Padua," *Journal of the History of Ideas,* I (1940), 177–206. As for Gilbert, who was a physician and indeed royal physician to Queen Elizabeth, and who had studied medicine in Italy and possibly at Padua, the medical influence in his *De magnete* may be discernible both in the insistence on empirical observation and experiment and in his so-called animism, which is essentially an insistence on the functional relatedness of living things and earth and sun and stars.

and the bedside are places where the knowledges of the other places must be focused. Medicine is then a kind of working philosophy of science.

Classical works on medicine are eloquent concerning this humane focus of the sciences. They reflect the scientific atmospheres of the times in which they were written, and at the same time they rise above these to see the whole human world within which and for which the sciences exist. It is not an accident that the Latin word for teacher, *doctor,* is used most especially for the physician, nor that the physician is pictured most often in drama as the knower of men.

It is my advice that these doctors ought to be consulted as the great conversation starts for the common reader, and that they ought to be consulted from time to time in the course of his reading. They will provide encouragement for what is really a great adventure, and they will act as correctives, regulatives, and antidotes for the *hybris,* or pride of the intellect, which is the besetting sin of liberal artists. It should not be forgotten that the Delphic oracle, which still presides over the affairs of reason, engraved two injunctions on its stone portals, *Know thyself* and *Nothing too much*. These are also the medical doctor's general prescriptions.

Hippocrates, now a legendary name not only of an individual but of any leader in the medical guild in the Greek world, is the name of a body of literature, the Hippocratic writings which come from several stages of Greek scientific thought. They are a record of its application to Greek men. *Air, Waters, and Places, The Sacred Disease,* and *Ancient Medicine,* together with *Prognostics* and *Epidemics,* are a good beginning sample.

Galen, who gave his name and influence to medicine for fifteen hundred years, perhaps too long, is a better mind than his well-worn reputation would indicate. He wrote many works, but *On the Natural Faculties* is recommended as speaking powerfully of both the substance and method of medicine, and therefore of the schools of science that flourished and fought for more than half of the period of Western culture.

Claude Bernard is still the most luminous figure in modern medicine. He almost achieves the single vision that is so much needed by both doctor and patient in the modern world, and where he fails the contemporary researches start. The *Introduction to Experimental Medicine* is the best introduction to the modern scientific world.

To the reader: Let the poet and the physician in you look over your shoulder as you read and discuss the great books of science.

Part I. The Parts of Science

IN THE long history of science there seems to have been a deep agreement that any science has three parts, variously named, but constantly distinguished, even by those who would like to increase or decrease the number of divisions. There is always something abstract and relatively fixed; such things are most often called principles. There is always something supposed or recognized as underlying the superficial or artificial stuff that the scientist is investigating; such things are generally called hypotheses. Finally, there are facts observed or made, opinions about such facts, operations performed on them by man or by nature, events, appearances, sense data, and even feelings. This third miscellaneous part has often been called experience and nature, but I shall want to call it the superficial and artificial part of science, and move freely back and forth across the line of the battle that experience and nature continually wage. I shall mean by *surface* the surface that nature presents to men, and by *artifice* the human activity that is rallied in any experience and meets and represents nature. I hope to show that human art has a large role in science; the word artifice will help. If the reader finds the words artificial and superficial inducing in him a contempt for this part of science, I beg his pardon and also his resistance to a bad habit of finding these words frivolous. All of modern industry is artificial, and Shakespeare is our greatest modern poet because he provides reality with an adequate surface. If science employs art to hold a mirror up to nature's surface, I want to make a just report of the difference it makes.

The other parts of science that are not mentioned here, and there are many, can be found as parts of these parts. It will be my aim to find some of them. I have chosen these three because I believe they are essential and comprehensive. One evidence for this is that most of the common-sense and philosophical interpretations of science consist in arranging and weighing these parts.

Principles

Principles appear in modern science as the last line of defense in a retreat before an array of findings that contradict or confuse a hypothesis. Physicists, habituated for a hundred years by custom more than by thought to the hypothesis of the luminiferous ether, an unbelievably thin but also unbelievably rigid fluid in which all physical objects swam, made great progress in investigation and increasingly great nonsense in thought up to 1905. I suppose that one of the greatest theories of all time, the electromagnetic theory of light, would have been impossible without the ethereal hypothesis. Nevertheless, there were curiously bothersome facts of observation and measurement that were accumulating without benefit of correction or explanation. Michelson and Morley imagined an experiment that would use their light-measuring gadgets and detect any motion the earth might have with respect to the ether ocean. They tried it and failed to detect the theoretically predicted evidence of the so-called ether drift.[1] The experiment was so simple and indeed so daring in conception that it managed to disturb the system of the physicist's knowledge as the other incorrigible facts had not succeeded in doing.[2] At this point the mathematical physicists retreated to the elementary notions of measurement. Einstein and others following him attacked Isaac Newton's assumption of absolute space and time; in terms of battle lines this was a treasonable attack, knocking out a supposed cornerstone of the structure of modern

[1] A small effect, much less than that implied by the assumption of motion with respect to a stationary ether, was in fact detected by Michelson and again in D. C. Miller's many repetitions of the Michelson-Morley experiment. According to a recent study led by R. S. Shankland, this effect was due to uncontrolled temperature differences (see Gerald Holton, "Einstein, Michelson, and the 'Crucial' Experiment," *Isis,* LX (1969), 186, n. 153).

[2] Buchanan is here following the account of the genesis of Einstein's relativity theory that was standard up to the 1950s and is still echoed in most textbooks of physics. From statements that Einstein made toward the end of his life, it has become apparent that this account must be revised. The Michelson-Morley experiment did not figure explicitly or crucially in Einstein's thinking prior to the publication of his 1905 paper, his preoccupation having rather been to eliminate certain fundamental assymmetries from electromagnetic theory. For a marshalling of the evidence and a plausible interpretation of this piece of history, see the article by Gerald Holton cited in the preceding note. The correction here required in Buchanan's text does not invalidate his description of principles as "the last line of defense in a retreat before an array of findings that contradict or confuse a hypothesis." In the case of the retreat to the relativity principle, however, some of the confusing and contradictory findings were the result of thought experiments—experiments like that of imagining oneself to be moving along with the crest of a wave of light—rather than actual physical experiments.

physics. Actually, the manner of attack discovered a deeper principle which not only provided support for the troublesome facts, but also suggested new and valid techniques of measurement, and most important of all, indicated the possibility of a new superhypothesis that would unify the theories of gravity and electromagnetism.

The deeper principle thus rediscovered in physics was called the principle of relativity. The strict mathematical formulation of it is simple in itself, but its interpretation and use in both mathematics and physics require a great deal of technical knowledge and know-how. In common sense and philosophy it is a very familiar principle: each thing is related to every other thing. In physics this has many special applications, recently in the study of fields of energy or systems of motion in which particles or parts of parts must be related to the rest of the system: all relative motions of electrons must be judged in relation to the speed of light.

The principle of relativity is a principle of modern physics because it is not simply supposed like a hypothesis, but it is presupposed, already supposed in any supposition we make. It thus rules not only in our investigation of facts, but also in our making of hypotheses to take care of the facts. It acts as first principle in the field of our investigation; without it we don't know where we are going or what we are looking for. With it we do know what we are looking for: if space is a network of relations, as the principle of relativity in physics says, then physics will be concerned with correlations.

The fomulation of the principle of relativity was a revolution in physics, and it is interesting to see what the revolution did to the previously ruling principles. Two of these were the principle of the conservation of matter and the principle of the conservation of force. These had been discovered and formulated for physics two or three hundred years before.[3] The principle of relativity showed the possibility of trans-

[3] The principle of the conservation of matter was a premise in the atomic philosophy of antiquity (*ex nihilo nihil fit*), and reappeared in the seventeenth-century natural philosophy in the writings of Descartes, Gassendi, Newton, and others. The first formulation of what is now called the conservation of energy is often credited to Huygens; to Leibniz we owe the name *force vive* or *vis viva* for the conserved quantity, as well as a number of clear expositions of the principle (see, for example, his "Essay de dynamique sur les loix du mouvement," written probably about 1691, in C. I. Gerhardt, *G. W. Leibniz: Mathematische Schriften* (Halle, 1860), VI, 215–31). It has been argued that the principle as it applies to mechanics is an implicit theorem in Newton's Scholium to the Laws of Motion in his *Principia.* But it was in the nineteenth century that the scope of the principle as applying not only within mechanics but to all natural phenomena came to be understood. Thomas Young, in his *Lectures on Natural Philosophy,* published in 1807, first proposed the term *energy* as a substitute for *vis viva,* but for another

forming matter into energy and energy into matter. The principle of relativity gave the quantitative relations between matter and energy in such a way that each of the previous principles was contradicted flatly, but at the same time it reestablished them in relations to each other under a new principle, the conservation of energy, and made each of the older principles a special application, or a step in an orderly process of specification.

There are many revolutions in the history of science like this, some of them constitutional as this one is, reducing ruling principles to coordinate functions under a new principle, others eliminating old principles to make way for total reconstruction. In an older terminology and perspective, principles are first premises of demonstration or proof, and the revolution appears as a dialectical search for better premises. Socrates in the later Platonic dialogues, and Aristotle in the many treatises that amplify his *Physics,* report the original pioneering work. In this perspective the search for principles seems to be motivated by a passion for first truths as the foundation for science. In modern science there is a commonplace assumption that regularly takes the place of this search, namely, that science is concerned with the facts. This also is a principle embedded in procedures of modern investigation; its formulation is difficult and confused for reasons that will appear when we come to discuss the third part of science.

Hypotheses

Aristotle said that all science begins in wonder.[4] He was probably thinking of the frustration of curiosity that accompanies the recogni-

half century yet, in discussions leading to recognition of the full import of the principle, it was generally referred to by some such title as "the correlation and convertibility of the forces of nature." In 1852 and the years following, William Thomson (later Lord Kelvin) undertook the terminological clarification resulting in our present-day distinction between force and energy. During the second half of the century, then, energy became one of the two great conserved "realities," the other being matter. Finally, what Einstein's paper of 1905 implied was the combination of the two principles into one, the conservation of matter-and-energy. For a history of the development of the principle of energy-conservation up to the end of the nineteenth century, see John Theodore Merz, *A History of European Thought in the Nineteenth Century* (New York: Dover Publications, 1965), II, 95ff.

[4] More exactly, Aristotle's statement is that *philosophy,* defined as the science (*epistêmê*) that investigates first principles and causes, begins in wonder (*Metaphysics* I, 2, 982b11–28); as such, he urges, philosophy is the only liberal knowledge or science, since it is pursued for its own sake rather than for utility.

tion of the so-called wonders of nature. Sometimes this is due to sheer size and distance, sometimes to depth and obscurity, but often, and perhaps always, to the bafflement of the powers of our senses, and of our understanding. The frustration of human powers very often is an incitement to imagination, guessing, predicting, and further investigation. These tries, so frequently found to be errors, compound the frustration and incite to further imagination. The control of such witchery is partially accomplished by the art of hypothesis, the second part of science that I have named.

Hypothesis in its original meaning is a strategy of retreat. Nature presents us with too many interesting objects. Selection is necessary and a deeper perspective is the only way to provide a sense of order and the possibility of choice. We back off from the amazing trees to see the woods, and the result is that something seems to be placed behind the appearances. A map does this for a countryside; the imaginary withdrawal of the mapmaker to an ideal distance allows him to "see" shapes, forms, and objects which place the familiar detail in perspective. The aviator who flew over the English countryside during the war and saw the ancient Roman camp outlined in shades of green in the vegetation was practicing withdrawal of this sort. In many senses he was finding the foundations of England. The Greeks called such foundations hypotheses, things "placed under."

It is true and interesting that astronomy was the science which first discerned this part of science. By necessity, and as it were, by providence, man is placed at the right distance from the myriad phenomena of the sky so that he cannot miss seeing the constellations. The Great Bear, Orion, and the Milky Way are the foundations or hypotheses of the systems of the world within or on which even the contemporary astronomer swings his reflectors and refractors. Our own earth, whether it be the humble footstool of God or a planet in a minor nebula, is such a hypothesis.

But these literal retreats and withdrawals are only the beginnings and symbols of the vast hypothetical structures that help to control our imaginings. There is a time and a need for rally and return, and the point of rally and the method of return are provided by the hypotheses themselves. Lucretius reports how Democritus accepted and stood on the foundation of the early atomic hypothesis which had its first revelation in the drift of the stars in space. But Democritus with the help of some good mathematics formulated the properties of the atoms so that they became the seeds of all things. He returned from this formulation to redefine and rearrange the four elements, earth, air, fire, and water. The conquest in terms of the hypothesis was complete, with explanations of even dreams, disease, and love.

A cycle of the hypothetical arts is not really complete until a second cycle is begun. The hypothesis demands more data than those that produced the first wonder and confusion. Armed with a hypothesis the controlled imagination divines new wonders to be sought. Prediction, experiment, and test invade nature. Curiosity becomes inquiry and research in which the hypothesis rules as if it were a principle.

There seem to be many kinds of hypotheses throughout the history of science. Some of them that have been crucial are simply things, whether spotted in observation or concocted in imagination, added to similar things that are familiar and puzzling. The planet Neptune was such an entity before it was observed. Its supposition helped to explain some curious irregularities in the motions of the known planets, the regularity of which had been formulated as a part of Newton's system of universal gravitation.[5] The colors between the known colors of the spectrum, or the notes between the audibly distinguished tones of the musical continuum, are more commonplace cases of the general category whose most spectacular representative is the missing link in the Darwinian chain of evolution. Such hypotheses are real and powerful beyond any common-sense estimate of their importance. The part played by the antipodes for the ancients or medievals, or the Northwest Passage for sixteenth-century monarchs, merchants, and navigators is nothing less than the secret of a whole atmosphere of geography, and therefore of the politics and culture of the time. These are things that could be and have been observed, but at least temporarily they were hypotheses. At any given point in the development of science there will be genuine hypotheses of this sort, and at such a point it will be impossible to know whether they are quasi hallucinations of heroic discoverers, or whether they are occult entities teasing men off the path of truth. One cannot even be sure that time will tell. Any large scientific organization that keeps records has filed away curious facts and guesses of this sort that may become important as missing pieces of a puzzle that has been interesting or may become so later.

Next to this kind of hypothesis there are those that are so imagined or understood that they cannot be observed. Democritean atoms were invisible not only because of their size but because they were colorless; their colorlessness was an intrinsic part of their nature. Likewise the modern atom and its constituent particles are too small by calculation

[5] Two theoretical astronomers working independently, Urbain Jean Joseph Leverrier and John Couch Adams, postulated the existence of the planet now called Neptune in order to account for irregularities in the motion of the planet Uranus. The first observation of the planet was made on the basis of Leverrier's calculations at the Berlin Observatory in September, 1846. For a history of this discovery, see Morton Grosser, *The Discovery of Neptune* (Cambridge, Mass.: Harvard University Press, 1962).

to be reported by a ray of light. Plato in the *Timaeus* reports a Pythagorean atom which was merely geometrical, regular solids which were completely empty and absolutely transparent. Another Greek atomist, Anaxagoras, proposed a qualitative atom which by itself was too small to see or touch, but with others in a group contributed to a visible quality. Mr. Eddington's table through which your elbow might pass without resistance once in a million tries is a more sophisticated example.[6]

It is usual to divide these hypotheses into two kinds in terms of connections they have with observations. Some of the hypothetical entities that cannot by themselves be observed, if they existed, together would have certain observable effects. They therefore would make a difference in experience. Certain others would not have any observable effects and would make no difference if they did or did not exist. The grace of God, ectoplasm or the soul, and Freudian wishes are sometimes blackballed by scientific societies because they have no consequences. This seems to be unjust to the hypothetical society; it is arguing from a cosmic provincialism that is itself often inconsequential. As Mr. Faraday said to Mr. Gladstone concerning his dynamo, You may be able to tax it some day. Politicians and reformers sometimes suppress the inconsequential hypothesis of the equality of man. At this date it seems that only individual human differences have scientific interest; common humanity makes no scientific difference, and the intellectual soul is not in good scientific standing. It is said by a very celebrated spokesman for modern science, Bertrand Russell, that a principle or an hypothesis that would apply equally to everything would be nonsense. Unfortunately, what this says about the human mind, the speculative enterprise, and the future of science does have consequences.

Of course, the thing- or entity-hypothesis can be exemplified in all sorts of fantastic and ingenious myths. The constellations that become animals in the celestial zoo, or zodiac, have biographies, just as the pygmy gambler in the Clerk Maxwell gas chamber has an impressive betting record.[7] The Chinese dragon who is driven away by drums

[6] Eddington makes his famous comparison between the "substantial" table of ordinary experience and the table of oscillating electric charges as described by the physicist in the Introduction to his book, *The Nature of the Physical World* (New York: Macmillan Co., 1928).

[7] Buchanan's pygmy gambler with the impressive betting record seems to be a kind of personification of the statistical theory of gases, of which Maxwell was one of the principal founders. The pygmy also appears to be near cousin to the sorting demon that Maxwell himself describes, in order to illustrate a limitation on the second law of thermodynamics. The second law asserts that every naturally occurring process will be found to be irreversible, provided that we take into ac-

and fireworks just as he is about to eat the sun on the occasion of an eclipse is a proverbial member of the hypothetical society. He is dated —or at least, his disappearance is dated—by the invasion of Western methods of verification. But the westerners bring with them a whole factory of machines to take his place. Some of these machines have a familiar history in the industrial arts: levers, pumps, pendulums, springs, and magnets. Some of them are also fantastic and ingenious: molecules, atoms, kinetic gases, and perfect heat-engines. Some of these gadgets of the imagination do not strain our credulity, but others break our minds, for instance the ether recently deceased. In fact, the mechanical imagination easily runs to paradox and self-contradiction. This tendency is not limited to such paradoxes as the perpetual motion machine. The mind of the mathematical physicist continually and happily outstrips his imagination, and playfully takes to the dialectical art of resolving the paradoxes it has conjured up. The books that have attempted the popularization of science have even used this display of wit to tease the lay mind into the fun of confusion. But the original confusion which these books were sharing has apparently always been fruitful of new discoveries and new hypotheses in the laboratory. Several times in the last two hundred years of mechanical thinking and imagining, the hypotheses have fitted together enough for the mathematical physicist to announce a system of the world in which all the machines have been hooked together to make a factory; the automatic factory for human use is only a late derivative application of the world-machine.

The life-history or progress of a hypothesis on its way to become a theory or law or system is fairly regular. It starts up as a correlation of similar observations, jumps by various associative laws of imagination either to another more familiar set of observations or to an imagina-

count the entire "universe" that is affected by the process; the direction of change is always such as to equalize differences of level, temperature, electric potential, and similar differences. Maxwell at the end of his *Theory of Heat* imagines a little demon manning a trapdoor in a partition between two halves of a container of gas; the demon would allow only faster-than-average molecules to go through the trapdoor in one direction, and only slower-than-average molecules to go through in the other direction, so bringing it about that one half of the gas would grow warmer and the other half cooler, in violation of the second law. This fantasy points to the *statistical* character of the second law, and suggests that a being or device capable of sorting individual atoms or molecules could reverse the otherwise inexorable trend in natural processes whereby energy becomes unavailable for useful work. The suggestion has been the subject of theoretical examination in recent decades by Szilard, Brillouin, and others, the general conclusion being that the device would not work, or would absorb more energy than it made available. For a review of the discussion see Werner Ehrenberg, "Maxwell's Demon," *Scientific American,* CCXVII (November, 1967), pp. 103–10.

tive construction like those in myths, collects more observations suggested by these associations, loses some observations judged to be erroneous on further trial, and finally clarifies and simplifies itself by dropping off the mere associative wrappings. Scientific myths tend to become abstractions.

In certain schools of rationalism and humanism where the great hypothesis of historical progress and human perfection has guided thought and history for the last two or three hundred years, this tendency of myths to become abstractions has had the doubtful benefit of a general formulation and application to the whole course of human history. I think it must be clear that the hypothesis of historical progress is itself a mythological hypothesis, taking its place of honor beside the ancient myth of the Golden Age; in fact, it is related to that myth as counterpart. In the earlier myth men looked back to a time when understanding was relatively easy and complete, when in theological terms they were immediately and constantly aware of being the sons of God in their minds. Historical time was the disturbing unrolling and progressive dissipation of an earlier natural wisdom. There was even a suspicion that men by taking their fates into their own hands had willfully lost their birthright. The myth played a part with a difference as late as the eighteenth century, when Rousseau wrote an essay to show that human inequality and injustice had been the results of civilization and the practice of the arts, including the liberal arts which were then making their great achievements in empirical science. It is interesting to note that this myth had currency when it seemed clear that the world was getting worse. The counterpart myth of progress accepted the challenge and made virtue of the vice of human independence. All past time was comparatively primitive and miserable because man had not taken care of himself, and had allowed himself to be deceived by man-made myths including the myth of the Golden Age. Human reason and even imagination were to be embraced and exploited if the world was to get better.

But even the early formulations of the methods of these schools of rationalism and humanism contained warnings of the difficulties in the program of human self-help. They were all concerned with politics and the conditions the social community provided for the exercise of reason, some like Spinoza realizing that the single human reason was far from adequate and that a republican form of government must provide not only freedom but also friendship for the great enterprise. These rationalists wrote essays on the improvement of the human understanding, prescriptions of ascetic discipline by which the imagination could be purged, strengthened, and trained to be the handmaiden of reason. Recently, the experimental method has been proposed as an easier substitute for such discipline. The concern in these prescriptions is to free the reason for its proper exercise in the medium of abstractions. The imagination is recognized as the maker of fictions for which the reason appears to have special and fatal affinities. In the earlier periods the imagination was supposed to be a part of the memory, all parts of which were ordered to intellectual ends. In the Renaissance the imagination with its mythopoetic ingenuity seemed to have declared independence and seemed to be seeking its own manifold ends. That these ends were still touched with intellect was not doubted, but regarding the result of some of the new alliances there was some reason to be alarmed. These were the days of a new alchemy, a new astrology, a new necromancy known as spiritism, and the Mason and the Rosicrucian, as well as the days of a decayed and corrupted religion. The prostitution of the intellect by the imagination was more than probable.

The strategy proposed was the detection of abstractions in the senses and in the imagination, and the tactics were tricks of abstraction of concepts from single concrete cases. These concepts are of two kinds, the so-called common properties of more than one individual case, the kind of thing that led to the classification of plants and animals in botany and zoology, but also led to useless monsters of the imagination like the philosopher's stone and the gold of the alchemist; and concepts of another kind which showed that the relations between things could lead to chains of reasoning. The rationalists and humanists either condemned the first sort of concept as the misuse of reason to make more and worse monsters of thought or ignored them in their emphasis on the latter. There was a dangerous conspiracy between imagination and the concepts of common properties; there was a legitimate and edifying alliance between the relative concepts and the reason. There was a sort of working compromise in actual procedure and also in terminology; one might use the common name for purposes of identifying or constructing an hypothesis, but the hypothesis must immediately be translated into its relational counterpart. The great discipline for doing this trick was mathematics, particularly the new algebra.[8] In ordinary language of yesterday's science the rule was to

[8] Algebra is in part a set of calculational devices that had already been developed in antiquity, notably by the Babylonians, and which continued to be widely used for practical purposes over the centuries throughout much of the Mediterranean world. But it was with the work of François Vieta (1540–1603), Simon Stevin (1548–1620), and Réné Descartes (1596–1650) that algebra became a completely symbolic discipline, and came to be conceived as the fundamental instrument for the understanding of nature. It is this development which justifies the expression "the new algebra." See Jacob Klein, *Greek Mathematical Thought and the Origin of Algebra* (Cambridge, Mass.: M.I.T. Press, 1968).

conceive the mechanism of the thing. Certain concepts that were particularly hard to transform this way were degraded to the rank of secondary quality and assigned to the work of the lowly and deceitful senses; the concepts that yielded relational understanding in terms of mechanism were treasured and processed in mathematics and assigned to the intellect or reason as primary qualities, or better, clear and distinct ideas. It particularly amused these founders of modern thought to think about animals as machines, and to make mechanical toys that would imitate animal functions. They also invented and used many machines for the purposes of measurement.

Most of our myths of progress about mankind's toilsome journey from the dark world of myth, magic, and religion into the clear modern light of mathematical physics and scientific method had their origin in some theme of these essays to improve human understanding. I have referred to the story as a case of a great and illuminating hypothesis in history to suggest that it is a rather heavily dated hypothesis perhaps kept alive too long, but, more important, to give some background for the proposition that all science moves from myth to abstraction. Perhaps it is better to say that any hypothesis will have an entity attached to it that can be discarded; so simplified, it will be abstract, carry a common name to hold its identity, and exhibit a relational structure. It is through this combination of characteristics that it can offer the kind of intelligibility that we call explanation. I suppose a machine still has the highest charm for us that this intelligibility brings.

I imagine that the most powerful, and at the same time the easiest, hypothesis to grasp is the atomic hypothesis. It has a history as long as the history of Hellenic and Western history, and probably is as old as the oldest myths of our civilization. Its life seems to run in cycles, some of which we know something about. It has the three aspects which I have been discussing clearly marked and easily distinguished.

It always seems to start with a principle, incidentally the only principle that modern minds treat as self-evident; that is, it is never argued for, always assumed, and used even when it is not made explicit. It might be called the holomeric principle: that anything is a whole which can be divided into parts and parts of parts. As Lucretius presents the atomic hypothesis, there are also other principles which he invokes: nothing comes from nothing, or everything has a cause; nothing that really is ever was not or will ever cease to be (an early formulation of conservation, later applied to matter and energy). The real verve in the hypothesis comes however from fascination of analysis, the progressive reduction of wholes to parts. There have always been many ways of cutting up wholes, but the most literal and obvious one is spatial division, as we do it with knives, saws, and chisels, or as the modern cement grinding machine does it to dust. After all the mechanical instruments are used, the imagination can continue and habitually does so until the principle of infinity is encountered. The mathematical habit usually embraces infinity and its consequences, infinite divisibility and infinite parts, catching itself only with a device like the calculus which keeps clear of the paradoxes of things infinitely small by stopping division where it pleases.[9] The nonmathematical mind stops short of the infinite and apparently postulates, occasionally with argument and proof, parts that are no longer divisible, are uncuttable, atomic. These uncuttable parts of wholes are the elements, the building blocks, the letters out of which the wholes are made by combination.[10] Here the myth-making imagination recognizes material entity, the atom. The hypothetical retreat from the appearances has been made by analysis, mathematical imagination, and then finally by a comparison with a homely concrete thing like a bin of grain; the atoms are the seeds of things.

During the first *floruit* of atomism as far as we know it, the exploration of the possibilities of rally and attack on the world of appearances was thoroughly made. There was the possibility that the atoms were of like nature with their wholes; this resulted in the qualitative atoms of Anaxagoras and their infinite kinds. There was the possibility that they were of very different nature from their wholes; these were the atoms of Democritus with their finite number of kinds. There were also the mathematical atoms, some numberlike, and some like geometrical figures. There were even atomic spirits or souls. This variety corresponds with the different styles of analysis. The modes of combination also correspond with the analytical styles and the kinds of atoms.

Recombination is the method of explanation, but before that can proceed with security, some notion of the properties of the atoms must

[9] Buchanan is here echoing a locution often used in the calculus to explicate the mathematical concept of limit. Thus one can say that an infinite sequence of numbers $a_1, a_2, a_3, \ldots$ has a limit L under the following conditions: any number ϵ being chosen, "as small as you please," a whole number r can be found such that for $n > r$ the difference between a_n and L is less than ϵ. This mode of definition emerged from the work of Augustin Cauchy in the 1820s, and was devised to free mathematics from the self-contradictory Leibnizian infinitesimals, originally conceived and manipulated as nonzero quantities that were at the same time infinitely small.

[10] Buchanan's metaphor is based on the fact that the Greek word *stoicheia* was used both for the letters of the alphabet and for the material elements hypothesized by various Greek philosophers.

be formulated. In the case of Democritus, according to the account given in Lucretius's *De Rerum Natura,* the atoms vary in size and shape. Sizes give the species or kinds of atoms, and therefore the common properties within classes; hence the generic concepts in the hypothesis. Shape determines the combinations or the possible relations that parts may have in wholes; hence the relative concepts in the hypothesis. As I have suggested, the Democritean atom was a very successful hypothesis over a wide range of observation in the ancient world. Man saw wholes made of indivisible parts; they built things they saw out of building blocks; and they spelled out the significant world from the meaningless letters of nature's book.

One gets the impression from the *De Rerum Natura* that there was more poetic imagination in Lucretius and possibly in Democritus than there was abstract science. Many of Lucretius's explanations are fanciful and calculated to make one love rather than understand nature. When the hypothesis was revived in the seventeenth century there was more concern about the mathematical structures and much less concern about probable sizes. It was realized that some atoms can combine with some others, and some cannot. Weights became important and the combining ratios between weights. Finally, invoking good arithmetic and a rule of simplicity, Dalton discovered that chemical combinations could be represented as combinations of atoms of different kinds, with small numbers of atoms joining to form the compound particle, and that the proposal had some striking and verifiable consequences.[11] If it were not for the myth of the atomic entity and its suggestiveness, the real atom could have been forgotten. Fortunately, the myth was kept; Dalton even drew pictures, although he himself thought of them merely as symbolic notations for the abstract arithmetic of his calculation. His equations had two meanings, one that would give the number or some multiple of the number of atoms in a substance, the other giving the weights of these substances that would combine in a new chemical compound. The first meaning served the myth; the second was of experimental interest.

Nevertheless, the next few steps in the use of the hypothesis exploded

[11] The most striking consequence had to do with cases in which one chemical substance *A* combines with another B under different conditions to form *at least two different compounds.* Then it turns out that the weights of *B* combining with a fixed weight of *A* to form the two compounds bear to one another a small whole number ratio, 1 : 2, 2 : 3, 3 : 5, etc., as implied by Dalton's assumption that in chemical combinations generally the number of atoms combining to form one of the compound particles is small, one atom of *A* joining with one or two atoms of *B,* or one of *B* with two of *A,* etc. The empirical fact here came to be called the law of multiple proportions.

the myth.[12] The analytic imagination and the experimental paradoxes together split the atom. As the molecule was the proximate whole which consisted of atomic parts, so now the atom was also a whole of which electrically charged particles were the parts. As valence bonds holding the constituent atoms together in the molecule were the net essential scientific results from the molecular theory of the time, so now the structure of the atom was to lead to a formulation of the relations between the subatomic particles. The splitting of an entity, absolute and impenetrable as atoms were supposed to be, revealed relations which could be expressed in arithmetical formulae. The findings concerning atomic weight were with slight revisions interpretable in terms of atomic numbers expressing the old ratios of chemical combination and also suggesting the internal mechanics of the atom. It is true that the atomic assumption still persists in the notion that electrons and protons are ultimate building blocks—if not indivisible, then as yet undivided. It is also true that the atom remains as a thing, and that other things, electrons and protons, resulted from the split. But the splitting goes on without limit in principle, and in fact, if we follow it through the science of radiation and the siege of the proton, culminates in the recent dissolution of the elementary unit of mass into energy. The science that goes with these developments gives apparently overwhelming evidence that the entities, that is, the material bodies of the original hypothesis of atoms, have dissolved into relations, thus adding plausibility to the doctrine that entities move from myths to abstractions.

On the other hand, our information concerning the life history, or life histories, of this hypothesis is not well rounded in the cycles or periods that we know. It appears that Democritus and Lucretius, or rather Democritus as we know him in Lucretius's account, was more

[12] The "next few steps" have to be taken with seven-league boots, the first step covering some fifty years of chemical analysis and argument leading to Cannizzaro's clarification of the system of atomic weights in 1859, and at about the same time to the elucidation of the principles of the structure of carbon compounds by Kekulé, Couper, Butlerov, and others. For succinct accounts of these developments, see L. K. Nash, *The Atomic-Molecular Theory,* in *Harvard Case Histories in Experimental Science* (Cambridge, Mass.: Harvard University Press, 1967) and Theodore O. Benfey, *From Vital Force to Structural Formula* (Boston: Houghton Mifflin Co., 1964). A further step was the discovery of Lothar Meyer and Mendeléyev in the 1860s of the Periodic Law, asserting a periodic repetition of properties as one runs through the elements in the order of their atomic weights; the law suggests that the atoms must have complex structures capable of manifold similarities and differences. The further steps leading to detailed models of the structure of atoms were taken principally by physicists, and depended importantly on Becquerel's discovery of radioactivity in 1896, J. J. Thomson's discovery of the electron in 1897, and Planck's introduction of the quantum of action in 1900.

concerned with the entities and their properties than he was in their shapes and relations; there is in other accounts indirect evidence that Democritus knew and used a great deal of mathematics in the development and application of his reasoning. It also seems at first sight that the seventeenth- and eighteenth-century atomists were more concerned about the relations and the mathematics than they were with the entities themselves; but here again it is clear that the entities were never dropped, and are not dropped at present. Perhaps it is best to drop the genetic account and note the three aspects of the hypothesis, the entities, the common properties with the corresponding generic concepts, and the relational structures or mathematical formulae.

Of course, there are other great hypotheses that do not follow the atomic and strictly mechanical style. Atoms have played some part in the theory of light, but on the whole a quite different kind of imagination and reasoning has determined its development. Optics has started from, and after detour has always returned to, pure geometrical hypotheses. The inverse square formula which was Newton's great discovery in connection with gravitation stands as a strong bulwark against multiplying nongeometrical entities. It is interesting to note that it applies to the propagation of light. Light from a near-point source falls on surfaces at different distances with an intensity that corresponds with the inverse square of the distance of the surface from the source. Edgar Allan Poe was so impressed with the inverse square law that he proposed it as the king hypothesis of all science. His essay "Eureka" may still have its contribution to make in the unification of science.[13]

Hypotheses contain imaginative entities; they abstract common properties and formulate concepts to correspond, and they exhibit relational structures at least in their mathematical formulations. Imagination seems to create entities, reason abstracts, and mathematics constructs as well as discovers order. Hypotheses have artistic, imaginative, speculative, and practical frames of reference. Perhaps all of these can best be seen together in some consideration of the oldest and most frequent expression for the function of hypotheses—they are the means for "saving the appearances." This is Plato's expression, a highly poetic and compressed expression of many of his insights about the intellectual

[13] Poe's long essay of 1848 has many remarkable features including an anticipation of the big-bang theory of cosmogony, and the proposal that periods of explosive expansion alternate with periods of gravitational collapse, as assumed in present-day accounts of the origins of the heavier elements and the solar system. A fundamental and perhaps romantic premise of the argument is that the results of physical science should be consistent with a human intuition that Poe asserts to be profound and all-pervasive—the intuition that all things must have emerged from unity and must finally return to it.

enterprise. Few readers of Plato seem to recognize how much Plato loved appearances, how in fact, like Shakespeare, he was the great artist of the superficial. It would not be too much to say that his whole effort in the dialogues is to understand and strengthen the hold that the Greek mind had on the world of natural change. The imaginative and speculative brilliance of the Greeks was endangering the integrity of their primary and elementary grasp of things. Plato made it his business to organize the imaginative and speculative findings so that the appearances could be saved. In this the hypothesis was the major strategic device. The hypothesis could make sense of the facts and thus could save them from unintelligibility.

But this function of hypotheses can obviously be discussed only as we come to terms with the third part of science, the stuff, the data, the flux of experience and the surface of nature.

The Greeks very early identified the defining characteristic of this stuff: it is whatever changes. Their last words were also closely related to this: nature is whatever changes. Physics, the science of nature, is the science of change. But change is not an easy, self-revealing subject matter for enquiry, and the Greek discussion of it ranges all the way from finding change self-contradictory and therefore impossible and nonexistent, to the famous demonstration in Aristotle's *Physics* that change is real, that it does exist.[14] As the argument that showed the impossibility of change concerned itself with principles, so Aristotle's argument that it is real concerns itself with principles. In fact, the principles of change are for Aristotle the principles of physics. I suggested earlier that principles are not always found in the heights and depths of thought, but often in experience and concrete facts. This is the great case. Aristotle established the reality of change by the formulation of its principles. Many sophisticated scientists would now say that his principles are inadequate to deal with the phenomena and some would add that they are vicious, but the answer to them seems to be that Aristotle's formulation has taken such deep root in language and in common sense that his critics are unaware of the use they are making of it in their criticisms of it as well as in their habitual practice in the laboratory and in their studies.

Aristotle observes that there are six different kinds of change in

[14] According to Aristotle, the existence of motion is obvious by induction, and in any case the physicist has to assume the existence of change since change is his subject matter, the nature or *physis* of a thing being defined as a principle of change or motion (*Physics* I.2, 185a12–14). As for the arguments of Parmenides and Melissus to the contrary, these rest on false premises, and their conclusions do not follow (see particularly *Physics* I.3, 186a16ff.). Zeno's arguments against the possibility of motion are also refutable (*Physics* VI, 233a13–32, 239b5–9).

nature, change of place in locomotion, change of quality in alteration, change of quantity in the increase and decrease of magnitude, and change of substance in generation and corruption. Although these categories of change are radically different and each contains many varieties, each observed case shows the initial presence of a form and then its replacement by a contrary form which was not originally present. The minimal assumption is the presence and absence of form, or, as Aristotle puts it, form and privation. These are two of the principles of change. It is interesting to note that they come to light in the mere description of the appearances, as principles often do.

But it should also be noted that a great deal of highly articulated analytical thought is expressed in the terms of the description. A great deal of the previous puzzlement and argument about the mysterious appearances has been ignored and the remainder has been employed with great care to avoid paradox and contradiction. Some of these troubles are recognized and brought back in the third principle of change. In change that has been described as transformation, or metamorphosis, as Aristotle has done, there must be a substratum underlying and continuing under the process, and this underlying something must have the power to possess and to deprive itself of the forms. It is in this notion of a substratum with the potentiality to possess contrary forms at different times that Aristotle accepts and resolves all the paradoxes of change. It is a conception that has puzzled and dominated the whole enterprise of scientific investigation since the Greeks. At the same time it has been apparently built permanently into our understanding of the world.

For example, a stick of green wood is thrown on the fire. Cold and wet, the noticeable qualities or forms of the stick, are replaced by hot and dry. It is not true to say that cold has become hot or that wet has become dry. Something that was, or had, cold now is, or has, hot; and likewise with wet and dry. Something that had the power to be either cold and wet, or hot and dry, has now demonstrated its potentiality with the help of fire. The description of the appearances and the attribution of them to the substratum become an explanation if we introduce the hypothesis that wood is the proximate matter of the stick and that it has the properties, or potencies, of absorbing moisture and heat, as well as of combining with oxygen in combustion. It is very much easier for us moderns to use the language and conceptions of the hypothesis and to forget the principle, but it is doubtful if we would ever have had the hypothesis if the principle of the substratum had not been discovered and formulated by the Greeks.

In fact, the substratum is the principle of all hypotheses, and this can be shown in more than one way. Superficially and linguistically, the

Greek word for *substratum* is merely a grammatical variant on the word *hypothesis;* they both mean "what is put under" and "lies under" as a foundation. We tend now to think of a hypothesis as something that we find or put back of appearances, and to think of the substratum as something that underlies change, but if we recognize that phenomena are the appearances of natural change, hypothesis and substratum should be the same. Actually, Aristotle often uses the Greek word for matter in place of substratum, and this usage throws considerable light on the place and function of hypotheses in the historic development of science.[15]

Matter, in Aristotelian metaphysics, is the most general term for all substrata, for that which lies under or is subject to all forms. As pure or prime matter, it is the refinement of a limiting concept in a universe which was first imagined as consisting of things made by a divine artistry subjecting matter to forms. In the natural sciences it appears in the various hypothetical substrata of change, identified on occasion with earth, air or water, most often with bodies of one kind or another, and under mathematical analysis with the ghosts of bodies as they appear in complexes of space, time, and motion. These are the entity hypotheses, but each of them provides some pattern of order that saves, or makes sense of, some of the appearances. It is in this way that matter, the all-pervasive substratum of all change and the great hypothesis of hypotheses, saves all appearances.

Facts

THE stuff of science—the so-called facts—has this puzzling context in change. But the facts have other characteristics. The phrase *matter-of-fact* indicates one of these. The fact is there whether you like or not, whether you helped put it there or not, and it will remain a fact no matter what you do about it. The fact is something done, and done with; it is stubborn, brute.[16] It has an independent concrete being, in spite of the truth that it might not have come into existence. But connected with this there is another paradox: you can be uncertain about a fact. Facts appear to be what they are not, or turn out to be something other than you thought they were. They are highly illusory. They can

[15] The Greek word for *matter* was *hylē,* which had earlier on meant "wood" or "timber"; the word for *substratum* was *hypokeimenon,* literally "what lies under."

[16] The word *fact* is derived from *factum,* the past participle of the Latin verb *facere,* "to do or make."

be stubbornly illusory. The facts seem incorrigible; we cannot get away from them, and we cannot possess them with security.

The ancients, perhaps because of their historic distance from us, seem not to have been worried about this puzzle as much as we have been in the last three hundred years. They even seem to have a masterful way of dealing with it, although their solutions do not meet our needs; they retreat too easily. Facts attack and invade the senses. After their first onslaught they remain in the memory as shadows and symbols. The real things of perception that we dote on are for the ancients merely the passing subject matter of passing opinions. If we try to pin them down or put our hands on them in the birdcage of our minds, they flutter away or die in panic.[17] Another figure of speech that is often repeated in the ancient literature describes factual things as mechanical toys which we set going and which walk or roll away from us even while we watch them.[18] The favorite word used to designate them is *phantasm,* and our modern word *fantasy* is only a slight coloring of the original borrowed from the association we make with poetry and illusion. The ancients were not deeply worried by phantasms because their quest for science and certainty moved through this medium to hypotheses and principles. A mind such as Plato's could even be happy and grateful for the suggestions and intimations that his images and idols afforded. The texture in which his mind moved can be seen in the Allegory of the Cave in the *Republic* and in the *Theaetetus,* which in some incredible way anticipates all of the last three hundred years of epistemology or theory of knowledge that we have just worried through.

I think we can begin to see the source of our unhappy enchantment in the two much-loved characters in Cervantes's great study of illusion. Don Quixote himself was the masterful savior of appearances, the fertile maker of hypotheses, the man who acted on principle. Sancho Panza tried to possess his own experience; he hoarded his sensations in deeds and facts. At the same time, he trusted his master, and was worried by him into real and complicated illusion. Modern man is both the Don and Sancho; his spiritual existence depends on his capacity to accept illusion deliberately and knowingly. This is nowhere truer than in science.

But this is dark speech. The plain account that drops some of the real mystery is simple enough. Our simple immediate personal experience carries with it an unassailable certainty. We cannot escape it, we

[17] The image of the birdcage is introduced in Plato's dialogue *Theaetetus* (197c et seq.) in the course of a discussion that seeks to discover the nature of knowledge.

[18] For this figure of speech, see Plato, *Meno,* 97d–e, and *Euthyphro* 11b–c, 15b.

do have it; in fact, we are a part of it. As the scholastic proverb goes, we make no mistakes in our sensations. But this overwhelming sense of certainty and inevitability, the tang of life that is in the epistemologically famous taste of pineapple, the sensed color yellow, the vision of the circling planets of Jupiter in Galileo's telescope, and the fountain of feeling connected with human love, these experiences are beyond doubt. They raise the question of certainty only about all other things. They exist, and the contemporary existentialism, like a certain kind of empiricist in science, makes the most of it. The fact is that we have sensations. The question of what they are, or what we do about them, is full of uncertainty and confusion.

The modern epistemologist has tried to plumb the depths of this puzzle by constructing the counter-character to the angel. An angel in our tradition is a pure intellect with no senses. He possesses no senses, but because of this defect, if you please, he does not need, nor can he exercise, any discursive reason. He grasps immediately and intuitively pure ideas or essences. Imagine, on the contrary, says the modern psychologist, a being whose only power to know is in his senses; suppose, although contrary to fact, that a newborn baby is such a being. The whole physical universe glares and blasts at him with stimuli. His sensory and psychic apparatus responds. What will be his experience? William James, with an almost angelic insight, says that the baby's experience would be a "blooming, buzzing confusion." Sigmund Freud, with rather more intense sympathy, says that the human psyche is originally very nearly in this state, and adds that the psyche suffers many wounds, imagined as burns and bruises, and that the first reactions of the psychic and sensory mechanisms result in the crowding out of some reactions to allow for others, as well as the anesthetizing of certain sensory areas, in order to save the organism. No one can doubt the certainty or fatal necessity of the knowledge that remains in the baby, but it is also obvious that further knowledge, even perceptual knowledge, is precarious. It is no wonder that we get concerned about education of the young and the progress of science. We may be forgiven if we wish we and our children were angels.

It is the confusion of this extreme certainty or fatality of factual knowledge with the extreme precariousness of any inference or conclusion from it that keeps us chasing, refuting, and re-creating hypotheses much to the embarrassment of any attempt to formulate scientific method. Immanuel Kant puts the case very clearly in his *Prolegomena to Any Future Metaphysics:* All metaphysicians are put on notice to suspend all their labors as futile until they have shown how the judgments of experience, that we have sensations, are validly related to the judgments of perception, that the objects of perception exist. It is all too true that the labors of metaphysicians have been suspended from that time till now, because we have not been able to answer the question.

I have suggested that there are two places where the scientists and we, the common men, meet the stuff of science: the surfaces of nature, and the artifices of our reactions. Facts are the things that result when these two meet in our experience.

Aristotle has the clearest and most incisive report on the surfaces. In what has since become common sense, he recognized that in the real processes of nature there are things undergoing change, substances with accidents that come and go or with forms that appear and disappear. Some of these accidents, particularly qualities, are sensed by our sense organs, which have the power to peel off or abstract the qualities. The phantasms that result are singular species, the taste of pineapple, the sensed color yellow, and so forth; or this wet and this dry, this hot and this cold. Although they result from contact with individual things, they are abstract and universal, capable of occurring in connection with other things, times, and places. They are called in this tradition sensible species. There is an impressive continuity of agreement on this quasi-psychological point even to the beginning of this century. One of the latest names for the image that carries universality and thus functions as an abstraction is the generic image. Considering the intricate complexities of the problem, the common recognition of this essential point is remarkable and can carry us a rather long way into the series of problems that it raises.

I shall comment on only two of these problems now because these two will be enough to indicate incisively what the distinguishing traits of this third part of science are, and because they are crucial in the differentiation of two prevailing lines of method in modern empirical investigation.

Although Aristotle seems to have had the original capacity and training which made it possible for him to choose as well as discover the most strategic and fundamental insights both on the highest level of principle and the lowest level of perception, there are few scientific minds since his time which have had that power. One of the consequences of this habit of his is that he always limits himself to a finite number of objects and these objects have a finite number of relevant properties. It seems obvious to him, and therefore it is unnecessary for him to argue, that this must be the general condition for any successful scientific investigation: one must somehow be able to choose the right number of objects and the right number of characteristics of these objects. Few of his contemporaries and almost none of the good minds since his time have been able to meet these conditions without accepting the identical objects and properties that he pointed out and defined. Many of the revolts against Aristotelianism have opened their attack by refusing to follow him at this point.

I am joining such a revolution when I say that the individual things that science investigates are infinite in number and each thing has an infinite number of characteristics. On the other hand, it is true that these things and their properties must be brought into a unity of understanding in terms of a finite number of concepts so related that they make an intelligible system. This makes a choice of concepts and therefore a choice of the various kinds of analysis necessary—some finite number of concepts and presumably one kind of analysis out of a possible infinite variety. Nine out of ten people who would competently consider this question at present would say offhand that at any given stage of investigation there are present in the minds of the investigators a finite number of ideas and skills of analysis and that habit, temperament, and laboratory atmosphere will determine which actually gets put to work. This amounts to dodging the question, and in fact makes most of our scientific investigation uncritical and irresponsible. Furthermore, it is evidence of a whimsical childish will, or many such wills, dictating in the manner of petty tyrants the ways and means of research.

I have no answer, but I should like to state the question more clearly. If we were omniscient, not actually as God is supposed to be, but merely potentially omniscient in the sense that we had the power to see all possible modes of analysis and to see the consequent facts that would result from any given mode of analysis, we would not need to be confused in our investigations because the fact showed an apparently infinite number of characteristics; we would easily make choices between well-differentiated alternative methods and between categories of facts. Aristotelians still claim that Aristotelianism gives them that power, but there are and there have been very few strictly Aristotelian investigators. The case of Galileo and his work on falling bodies illustrates the point I am trying to make. As Galileo tells the story, he found the Aristotelians arguing as if from the authority of the master that bodies fall with a speed that is proportionate to their weight, the heavier body falling faster. In experiments that may have been performed from the top of the Leaning Tower of Pisa, Galileo found that there was little but an accidental difference between the speed of the heavier and the lighter bodies. By a piece of analysis that the Aristotelians had not even thought of, mostly pure geometry, Galileo became convinced that the acceleration of the freely falling body was constant so that its velocity was proportional to the time of fall, and moreover that this constant acceleration was the same for all bodies, no matter how heavy or how light. He tried some more experiments and found that the facts confirmed his hypothesis. Later, still more careful experiments in a vacuum and with an inclined plane supported and still support the hypothesis. The story is told in every history of science to show the daring originality of Galileo and to show the authoritarian stupidity of the Aristotelians.

But there is more in the story than this. The context for Aristotle's dictum that speed of moving bodies is proportional to weight contains more characteristics of the facts and also a quite different basis of analysis. In what might be called Aristotle's dynamic equations there appears a term representing resistance of the medium and it is obvious from this that Aristotle was interested in motions of falling bodies in various media. Experiments carried on by the Navy during the last war on the dynamics of depth charges falling in the water would be more relevant than the findings of the Leaning Tower. Aristotle was ruling out the possibility of a vacuum, and would have found the postulates of Galileo's laws of motion quite strange. It is pretty obvious from this addition to the story that the two investigators were using quite different methods of analysis and were viewing quite different aspects of the facts. It is quite true that the methods have some relation to each other, as it is also true that the *fact,* falling body, is common to the different thoughts about it, but I am not at all sure that Aristotle and Galileo would have gotten together even if they had had Newton in as a *tertium quid* to the discussion. I am not sure either that future physics will continue to rule Aristotle out and give the prize to Galileo. The issue is not between a right and wrong account of the facts, nor between care and neglect of either facts or ideas. The issue is in the facts themselves, which contain a greater variety of possible aspects, hence parts, hence methods of analysis, hence ideas and explanations than even Einstein with yet another slant on it dreamed of.

There is a similar case in the Galenic and Harveyan theories of the circulation of the blood. Galen saw the blood like an ocean moving tidally in and out of the organs; Harvey saw it circulating through pipes away from and back to the heart. Neither of them had seen the capillaries, and today it appears that blood in the capillaries comes and goes with tidal periodicity.

The point here is a commonplace of the courtroom where it is taken for granted that the facts of a crime, their relevance as evidence and their weight as proof, depend upon the legal presumptions, the precedents brought to bear, and the intent or spirit of the laws. Even a novelist knows that it takes an artist to know what actually happens even to fictional characters. Facts contain a manyness that outruns the analytical and observational devices, but the aspects that are caught and brought to recognition attain scientific status only by the articulation of abstractions that rule the various systems of our knowledge. Each of these systems falls far short of the divine omniscience that can contain the infinity that a fact presents; they also fall far short of the Aristotelian precision that can rigorously eliminate irrelevance. All this has been summarized in the proposition that an individual thing is the coincidence of an infinite set of abstractions. Although this may have some shadow of metaphysical meaning and validity, it doesn't help either the day-to-day investigator nor his responsible critic who would judge the authenticity of the facts and their explanation. Peel off all the abstractions that can be discerned in the fact, there still seems to be left a prime unformed matter in which a better mind or a later investigator could and probably will discern better and clearer forms and abstractions. The facts of the case are apparently the permanent possibility of more and more analysis.

A radical and extreme solution of the problem is presented by Kant in his *Prolegomena* where he is concerned to account exhaustively for the prior conditions of scientific facts. After finding the categories by which facts are judged in the kinds of propositions that can be made about them, and after adding to these the conditions in time and space under which observations are made, he announces with some assurance that the mind endowed with these categories and intuitions makes the nature which presents itself in our experience. Whatever else there may be we cannot know and could not be a fact. We have enough to do if we watch ourselves constructing the nature and the facts that we do know. This is not as perverse, subjective, and negatively skeptical as it sounds at first. It is pretty close to a description of the investigator's actual procedure. To understand it better we shall have to turn to another aspect of the facts.

It is pretty clear that whatever we may think of human life in general, the sciences are not attained without effort, without planning, without purpose. The scientist is a man with a purpose, and his efforts are directed by the purpose. The efforts require skills of all kinds, skill of hand, skill of eye, and skill of imagination and thought. In a very real sense he must bend all his experience to the end for which he works. That end is conventionally and correctly recognized as knowledge. He is successful when he has learned and taught something. Mere rules of thumb and routine, no matter how thorough, no matter how helpful, are not enough. Successful technology, which in modern times has given the popular glamour to science, is not enough. These are all means to the end of understanding.

In spite of this theoretical end, however, science as an activity is practical. The practice is for the sake of theory, but it is nonetheless essential in the whole enterprise. The scientist makes things and does things; he operates on material. It would be surprising if this operation did not leave some marks in the knowledge. It may be true that the kind of knowledge that would be the proper or perfect end of science would have no such marks of operations left in it; it might be that it would have no application except in contemplation. But as we know science, it has plenty of signs in it of the workman, the human artist. Not only has it marks of past operations, but it even seems to have signs in it of its possible future application. This is true of the other two parts of science, principles and hypotheses, but it is particularly true of facts, as the etymology of the word indicates.[19]

There are two generally known and discussed instances in which operations, the operations of investigation, destroy facts. The older and more familiar case is in the investigation of living organisms. Anatomy and physiology have at times in the past had to deal with dead bodies. The cutting, the seeing, and the experimentation necessitated the killing of the animal. Therefore, most of the facts of life or of vital function were eliminated at the outset. In this case the investigation destroyed the essential facts. The newer and less familiar case is in the study of the motions of the electron and other subatomic particles. Here the rays by which the observing apparatus penetrates the phenomena change the course and velocity of the electron. The extended fingers of the observer mess up the thing observed; the extended eye of the observer is an evil eye. The man stands in his own light.

But these are only special cases of a very general import. The hunter and the naturalist in the field have always had to be careful not to frighten the animals they seek, and the domesticators of animals and plants have always changed the habits of the objects of their care. The experimenter must always introduce artificial conditions into the experimental situation. Standard procedures and materials have to be substituted for random prodding and sampling. To make the apparent paradox acute, one may say that now random sampling and testing are themselves well-regulated standard procedures inside science.

By the most elementary logic it is obvious that the introduction of any operation knocks out at least one item of the situation, and some fact is not allowed to speak for itself. It is true that such operations uncover items and allow other facts to speak for themselves, but the picture of the blank mind allowing nature to impress itself without interference by the will of the operator is spoiled.

The complementary error in the determining of fact is so well known that it has attained a technical name, the *artifact*. It is disturbingly illustrated in the story of the investigator who had worked for three years on the measuring of the electrical forces in the innervation of a muscle

[19] See note 16 above.

in the frog's leg. He was a highly trained technician, had invented a special amplifying apparatus, was an honest, conscientious reporter of his work, and had a large mass of unique and surprising findings which no one else was able to get. In another year of confirmatory findings his work would revolutionize this lively subject in physiology. One day the electrician of the laboratory was hunting trouble in the circuits that supplied the laboratory and he traced an unmapped line that ran back of the wall board over the experimental bench where the frog's leg was making its remarkable record. From that day on the findings began to refute the three years' work, and it was found on further sleuthing that the circuit back of the wall board had been influencing the delicate amplifying apparatus and disguising the nerve currents in the frog's leg. This case is remarkable only in its simplicity. There are many cases like it in which the cause of the artifact has been harder to detect. In a world in which technology constitutes so much of our environment, and in which gravitational fields, electromagnetic fields, thermal fields are recognized as all-pervasive there is some cause to wonder about the effectiveness of screens and controls in isolating the phenomena under investigation. Man proposes and unknown to himself disposes many experimental routines.

Of course, the scientist is not unaware of this. In fact, a great part of the concern in setting up experiments or institutes of research is to know the environment, to vary the conditions, and to check the techniques for the artificial components. Young sciences that are breaking new ground and collecting first data actually have no protection from themselves in this respect. They must damn the artifacts as they appear and go ahead. Sciences heavily loaded with practical and perhaps moral overtones have to be doubly cautious because they may drift into occultism, which might be defined as the uncritical acceptance of accidentally manufactured data. Psychology stems in modern times from mesmerism and phrenology, and still has a bad time steering away from their modern forms. Race theory in biology as soon as it is applied to man runs easily into the occult. Field work with primitive peoples in anthropology has only recently become aware of the danger. But even physics and mathematics, older and more sober sciences, have fits of occultism. Kronecker, the mathematician, says that God made the integers, the whole numbers, and man made all the rest of mathematics. Eddington sums up a whole cycle of recent physics by saying that the physicists have been trying to guess the monster whose footprints they have been measuring; and behold, the footprints are their own. Kronecker and Eddington are reporting at least tentative and partial recovery from the inherent self-deception of scientific method. They are saying: I, a sailor shipwrecked on this shore, bid you set sail.

Of course, these two dangers to scientific navigation—complete extirpation of facts and complete substitution of artifacts—are extreme and perhaps merely imagined cases. What may be called the natural matter of the facts resists being put aside and registers its existence in that resistance. Likewise, the artifact is constructed out of the same kind of natural matter which presumably shows through the artifice. Extreme occultism as it occurs in travelers' tales and the imaginative stories that subtle natives tell naïve anthropologists are the results of deliberate attempts to substitute fiction for fact. The tougher and more delicate problem is also the more familiar where nature does not reveal itself without some aggressive guesswork and organized inquiry by operations. The cutting, staining, and fixing of tissues for microscopic observation is a good case. In order to get a slice of tissue that will let light pass through it the tissue must be impregnated with a substance like gelatine so that the microtome will cut cleanly. In order to make structure visible in the tissue it is necessary to introduce stains that will be absorbed in different degrees of intensity by different parts of the tissue. The stains are often dyes that form chemical combinations with the substances in the tissue. The transforming power of the dyes is suggested in the discovery that some of the dye substances are good bacteriocides in living tissues. All this preparation follows in most cases the death of the tissue. Extirpation and construction of fact seems overwhelming, and great care is needed to decide what, if anything, of the subject of inquiry is left that can be called a natural fact. But there is knowledge and there are relevant rules of operation by which such questions can be solved at least approximately. The physical and chemical effects of the foreign substances introduced can be ascertained. The knowledge of light and optics can be figured in and then subtracted. Independent knowledge of the tissue under investigation can be brought to bear. The residue from all this prodding, manipulating, transformation, and questioning is assumed to be matter of fact. This involves the peculiar process of reasoning from known to unknown, but the lack of rigor is or can be progressively remedied by later correction when more is known.

Many studies and reasonings in early modern physics are more daring with artificial devices. For instance, it is assumed in the theory of the pendulum that a pendulum has the mass of its weight concentrated at a mathematical point and that the supporting string or rod is a mathematical line with no weight. If this is assumed then the law of periodic motion of the pendulum, formulated by Galileo, makes sense of the facts, although no actual pendulum of the sort described could possibly exist. On the other hand, the law directs the setting up of the experiments and tells the observer what to look for. The observer reports degrees of variation of actual pendulums from the behavior of the ideal pendulum of the law. In this case the facts are marked by their variability from an artificial norm, and they would not be known at all without the artifice. Fictions of this sort are scattered all through science, and their vital role is most often the vantage point they give the observer to determine fact by eliminating irrelevant data.

In the case of the pendulum, the artifice that focuses and brings out the facts is mathematical, but it is closely related to the whole arsenal of machines, both material and ideal, that function partly as hypotheses and partly as fact by previous determinations. The hypothetical role that machines play has been mentioned earlier; but the machine as an artificial device injecting operational items into the stuff of observation is perhaps the best all-around illustration of standard practices in investigation. The Greek word from which our word *machine* is derived was regularly used by the Greeks to refer to a tool of some manual art.[20] As such it regularly connotes a means to the end that the art serves, and distinguishes recognized and regularly employed articles rather than chance conditions of artistic production. A tool was also itself the product of the art of toolmaking. We do not escape the implication of purpose when we try to make a machine automatic; it is still an embodied purpose, as was well recognized when mechanical science saw all of nature as God's machine.

The soulless automaton of the nineteenth century and of the industrial revolution is a deliberate and emphatic contradiction in terms, if not in reality. But this curious paradox is saved when we see what the investigating scientist does with the actual machines he uses. They are tools of investigation and their end is high, the determination of fact. A machine even of the simplest construction embodies mathematical principles and its application almost by magic transforms relatively indeterminate stuff into numbers. This is true of the balances as well as of the most recent calculating machines that make mathematical tables. It stretches things a bit to say that nature contains numbers, if by nature we mean the observed facts. It would be easier to say that the machine imposes numbers on nature. It is probably better to say that nature is potentially mathematical, and that the machine activates the mathematical principles in nature—*principia mathematica naturae,* as Newton called them. A machine is a reasonable device for persuading nature of its mathematical nature, and it operates gently and delicately to encourage nature to express that side of itself. The artifice, like many human artifices at their best, is to enable nature to know itself. There may be better devices than machines to do this, but so far they are the best

[20] The Greek word referred to is *mēchanē.*

and safest artifices that we know. They confer a kind of human reason on nature, making it intelligible, if not intelligent.

One more comparison, if it is not too farfetched, may help to get a little clearer the confusion of artifice in fact. The experimental scientist is like a playwright, and his laboratory is like a theater, a place for seeing as well as acting. He introduces artifice for the same reason that the playwright does, to make nature more eloquent. There is of course mere showmanship in both places, but the genuine performances in both places are aimed at demonstration, demonstration of the facts and showing their inner natures. Actors, both facts and hypotheses, are brought on the stage and made to behave in character, perhaps by learning new tricks in made-up situations. They may go through various antics that disguise themselves and deceive the spectator, but the play-acting is deception with a purpose, the demonstration of truth, and failing to achieve that purpose is recognized as failure. Different experimenters may have different styles, but in the end they allow their art to come before the judgment of its end, the discovery of truth. There is in fact a deep connection between good science and tragedy. The hypothesis that maintains its heroic role throughout the ordeal by experimentation and facing the consequences of its acts may find its fulfillment in sacrificing itself to the facts as they are. One of the important commonplaces of the laboratory is the recurrent discovery that the successful experiment is one in which the guiding hypothesis has been refuted.

But this is enough about this part of science. We are very far from being able to say what the facts are, even in principle. It is unavoidable, but irritating, to have to say so much about what they are not. Like the prime matter of the Aristotelians and the God of the theologians, they can properly appear in discourse only in things said about them by more or less indirection. I have talked about them in terms of sensation, of abstraction, and of operations on them. They are the stuff of these processes, and therefore of our experience. Perhaps a mathematical expression can be borrowed for a final word. The facts are "such that" they can be sensed, universalized, and made into knowledge. They are potentially intelligible. In the grand manner of the mathematical logician, temporarily turned metaphysician, the facts are everything that is the case. They are the material cause of the discourse.

But before we leave them for easier matters, I think I ought to report a word of warning from two recent philosophers, each of whom spent a great part of his life in a difficult branch of scientific investigation. They agree on the point I wish to make a warning. The facts with which science deals are not irreducible stubborn stuff of human experience and actuality. God, who sees everything, and for whom therefore everything is as our limited experience is to us, may possess the concrete certainty which we try to attribute to scientific fact. Whitehead says God is the principle of concreteness. We do serious damage to our knowledge when we misplace the concrete in our experience. We arrive at our facts of science by a difficult process of abstraction and feeling. In the depths of our private experience we may know a very few things as God knows them; one such all-pervasive fact is that we have experience. Bergson, the other philosopher with scientific experience, thinks that our mechanisms and our purposes are comic parodies of our actual knowledge, in which actuality is a spontaneous fountain of life grasped only momentarily as we transform it into the abstractions with which we play, sometimes suffering as children do. The existentialist who thinks he can recover all his experience in the depths of his private being is trying comically to be God. He might better recollect his emotion in tranquillity.

This concludes an all too brief and at the same time an all too wordy introduction to the main theme of this essay, namely science as the centerpiece of the human speculative enterprise. The brevity and the wordiness are due, aside from my own limitations, to the state of the subject matter. Modern science is a going enterprise with an unorganized center and fluctuating boundaries. It is aggressive, ebullient, and uncertain of itself. It seems to me that Kant gave the last adequate critical account of the enterprise for his time; since then the current has regularly outrun any mere report of progress, to say nothing of just criticism. There is a parade in modern science of which coming to be and passing away are the essence. Such phenomena, progressive, retrogressive, cyclical, or spiral, as they may seem, leave traces and precipitates. I have tried to indicate some of these. Such are the distinctions between the parts of science, and I note that I have called the distinguished parts by names that suggest the parts that Aristotle named for the science of his time. He called them principles, causes, and elements. I have called the parts that I see principles, hypotheses, and stuff. The change, I think, indicates our uncertainty and possibly our incapacity to see through our own constructions to the speculative realities. It may also indicate that Aristotle did not foresee the eventualities and puzzles which the continuation of his work would present in two thousand years.

Perhaps it will help to summarize what I have said under a frame of reference. In Aristotle's *Physics* it is said that scientific investigation begins with a confused whole of observation and opinion in common experience and moves to an intelligible whole of principle, cause, and element. We have some doubts about arriving at causes and elements, but we do work with principles regularly, if not explicitly. One of these principles is that anything susceptible to scientific study is a whole whose parts can be found by one or another kind of analysis, and when found they will be related to each other. This may sound too simple to deserve explicit statement, but I believe it is and always has been the correct description of the scientific enterprise. The concept of wholes with parts states the grand starting point, assumed if you will. Division or analysis proceeds by an oscillating attention to hypotheses and other parts, some of which are facts and some of which are constructions. The result is clarification of the analysis, or confusion and abandonment of the project. Hypotheses are of many kinds, facts and constructs are of many kinds, and each mixes with the others. Hypotheses get established as laws or theories, facts get established and accepted. Sometimes great systems of hypotheses and facts accumulate, sometimes they are destroyed by further knowledge or by substitution of simpler and more powerful hypotheses. Fatigue, failure of curiosity, moral confusion, politics may disrupt the process, but it revives. At present, we are facing a crisis in the internal organization of scientific research and its relation to other human enterprises. If the reorganization cannot be made, confusion and frustration may disrupt our going enterprise.

There have been long historic periods when the working enterprise has understood itself well enough to formulate its own methods. It may be well to give some account of these intervals of relatively clear self-understanding.

Part II. Methods

THE initial persuasiveness of Plato's doctrine that what a man learns or discovers is really something partially remembered from a previous life in which knowledge was complete, immediate, and certain, is due to the feeling of incompleteness that goes with all human knowledge. There have been times when men have been convinced that they had glimpsed the whole of which they and their knowledge were only parts. The glimpses of the whole have contained indications that if this or that were done, the missing parts could be found, grasped, and put together. The idea of a whole itself is enough to keep the appetite whetted, and it is only by a persistent ascetic discipline of the mind that new discoveries are kept from persuading us that they are the secrets of everything. The modern period in which the romantic imagination has teased us with mystery and the technical applications of science have constantly reminded us of magic has put us on guard against the sin of intellectual pride. But an equal and opposite pride has taken the place of the obsession with totality; we pride ourselves on our skepticism concerning wholes. One form of this pride is a general belief in progress, a progress that does not move toward a finite goal, but a step-by-step advance to more and better; our intellectual humility avoids the superlative and prefers the comparative. We see ourselves as wayfarers, and cure our wonder by wandering.

The style this disciplined appetite for totality takes in scientific investigation is a devotion to system and method. A system is a selected part of the great research viewed as if it were a whole, the concept of totality applied to a part. Carried to an extreme this trick of abstraction is the vice of specialization; practiced with skeptical fervor it makes the scientist a cautious, pedantic scholar; with critical moderation it measures the energies of the mind to the several arts that make the scientific enterprise. The essence of a system is the articulation of parts, and an assumption of a proper whole, arbitrary though it may be, helps by giving order to the procedure.

Systems dictate methods, or the rules for articulating parts and the operations that go with them. The last three centuries of science have been filled with proposed methods, methods of investigation, methods of training the mind for investigation. We now have a science called methodology, and there are methods of methodology. Each separate science has its own methodology, and there is a philosophical school that would reduce the whole enterprise of science to the methods by which it advances.

The notion that lies back of method, both historically and logically, is the notion of art, as the ancients understood it.[1] Arts are human activities directed to ends. The useful or industrial arts are directed to the making of things for use. The liberal arts are directed to the understanding of things. The divine arts are directed to the improvement and saving of souls. These are the large divisions of human activity, and it was the ideal of the early monasteries to supply the conditions for the practice of all these arts in proper proportions and organic relations by each man. Each art thus subsumed in the general scheme had its own respective end and its own special operations and skills. The acquisition of a skill was the achievement of a virtue. In the ancient world there seems to have been more concern about the possession of a virtue than about the external products. It may be that our modern concern about method is a return to the man and his skills from what has seemed in an intermediate period to be an almost exclusive concern about the external product.

Science, as the possession of a man, is a set of skills, habits, and virtues, vaguely and confusedly understood as methods. As each art has a corresponding science, the useful arts depending on natural science, the liberal arts on formal sciences such as logic and mathematics, the divine arts on ethics and theology, so methods go with systems and depend upon them for their systematic procedures. Sciences that have systematic articulation therefore dictate scientific methods, and there will be as many, at least as many, methods as there are sciences.

The number at present, and for a long time past, is myriad. There is not merely the array that might be expected from an orderly deployment of abstractions over various well-marked fields; there are auxiliary sciences for each of these. Then there are sciences such as biochemistry and astrophysics which combine others as their names indicate, their combinations involving the combining, if not mixing, of methods. These then multiply the original canonical number, if there be such. In addition, the different ages of a given science demand different strategies which amount to different methods. All this specification and multiplicity is enough to drive the systematic mind to extreme measures of codification in the science of methodology, and it is not surprising to find an avalanche of words wherever the attempt is made. It is

[1] The Greek word for *art* is *technē,* from which we derive *technique* and *technology*. A translation of it that Buchanan often used was "know-how."

not impossible to build towering edifices of synthetic philosophy out of this material, as Auguste Comte and Herbert Spencer did in the last century, but there is a suspicion of Babel hanging about the enterprise, even when agnosticism is included to neutralize any possible blasphemy.

The wiser plan seems to be to withdraw from the tower in the jungle, trusting that the workman on the job will make his own rules of thumb or of pseudo-metaphysics, and to note the depth and height of the effort in terms of the ends sought in the various fields. For in all method there is purpose and plan from which means and tools take their functions.

In the long view we are taking, it is fairly easy to discern the ends toward which methods mark the ways. The essential single aim of all science is speculative truth, and all method directly or indirectly leads to understanding. But however dominating the aim may be, its content or embodiment as a prize to be won is not single, and this is true even in theology, where the drive to know the one truth is paramount. Man does not understand God, as the doctrine of the names of God warns.[2] There must be many mansions even in that house of truth; there must even be many theologies if any is to be safe from falling into falsity and self-deception. Human knowledge is radically and inveterately heretical, in the sense that it takes the part truth for the whole, and it is the very beginning of wisdom to recognize this. So in science the one aim is achieved in many parts and stages, each defined by secondary aims, and in the mediums in which light is to be found. Attainable speculative truth is a dome of many-colored glass staining the white radiance of the one unattainable truth, but the glass transmits light, and its position and color can be ascertained.

The Three Methods

Each part of science sets its own end, claims and offers means and methods for its attainment, and tends to subordinate the other parts of science and their ends to its support and aid. Principles contain speculative truth, and they enlist hypotheses and facts for their discovery and clarification. Hypotheses move back and forth in a kind of dialectical dance for which principles supply the choreography, and facts the

[2] The doctrine of the names of God here referred to is sometimes called the analogy of being, and is set forth in Thomas Aquinas's *Summa Theologica,* Part I, Question 13.

music and costumes. Both principles and hypotheses appear and disappear at the beck and call of matter of fact. No method is wholly at the service of any one of these parts, but there are rather remarkable stages in the long, broad history of science and also in the narrower history of any one science, when the domination of one end over the others is clear and decisive. It is a matter of common report by the historians of science that the ancients from the early Egyptians and Babylonians to the scholastics of the Middle Ages were interested in principles. They were therefore content to accept common sense observation and the hearsay evidence of travelers' tales as fact, and they were particularly susceptible to myth and magic to supply their minds with hypothetical reasoning.

Although this account of the period appears to be itself a myth to those who have read the texts of Euclid, Apollonius, Archimedes, and Ptolemy or even such devotees of sharp observation and experiment as Hippocrates and Galen, there is a kind of truth in it for us moderns who like to view our predecessors as primitive originators of and contributors to our own intellectual enterprises. And it is indeed true that the ancients have given us a heritage of principles, some of which we honor more by implicit use than by explicit recognition. It is however also true that the sharpness of ancient vision and the richness of ancient imagination and memory are due in some degree to their discernment and formulation of principles which they then used as blueprints of a search. Essence and cause are two principles which were undoubtedly the ends of ancient method, but it must be admitted that as ends of research they lead and impel no slight labors of imagination and sight. It is true that the hypotheses and the facts that these labors produced belong to the more variable parts of science, and this truth may account for our low estimate of the historic findings. But with the connection that these remarks suggest, it still appears that the methods pursued by the early science of our civilization were aimed at the principles, and that the other parts of science were cultivated as aids and supports for that part.

The historian's account continues by indicating that the period we call the Renaissance marks a revolution against the high speculative end, and that principles and facts were used even intentionally to separate out, to abstract and to clarify hypotheses. It is as if principles and facts together became data for analysis, and out of that analysis almost automatically there came hypotheses, as clear and distinct ideas. There is a revival of ancient science in which the principles get applied to facts with new incisiveness and rapid success. The instruments, or organons, of method are the new algebraic forms of mathematics and the tools of industry inverted or perverted from their original functions in production to the arts of measurement. Hypotheses become mechanical, mathematical, and relational. Principles are dissolved in the formulae and the machines, and facts become the negative controls for sustained flights of imagination. There is a new and exciting beauty that shines out from the hypotheses that are read in depth from the book of nature.

As the ancients had used a high and subtle dialectic to come by their principles, so these minds that read the book of nature adjust and set in motion their algebraic symbols and their mechanical models. The symbols and the machines so faithfully scan the phenomena that they seem to be imitating angelic intelligences which not only know but direct the motions in the depths of nature. As Archimedes had asked for a fulcrum from which to move the world with a lever, so these minds ask for an axis of reference within which the semi-intelligent hypothesis, formula, or machine will almost spontaneously live and move. These are the living pumps in the human body, the soul of the magnet, the harmonies of the spheres, and the guided wanderings of the planets and the atoms. The clear and distinct ideas, discerned in the hypotheses, almost persuaded the scientists who cultivated them that nature itself was a society of angels in disguise, and there is a suspicion that neither the angels nor the scientist had need of either principles or facts, or if their weak imaginations called for such aids, the hypotheses themselves would provide them.

But there is in this story also a great deal of historical distortion. Although there is the high philosophical doctrine that a clear and distinct idea contains its own criterion and evidence of its truth or falsity, the writings of Copernicus, Galileo, Harvey, Gilbert, Kepler, and Newton contain much dialectical discussion of principle, and it is from their practice as well as from their critical reporting that we derive the modern laboratory and its methods. It is true that these men are the heroes of modern science because they constructed and concatenated the main hypothetical tissue of modern scientific thought, but to a remarkable degree they wove into that fabric the crucial factual data and the heritage of principle that gave their work the character of a renaissance. In their work, system, which depends so much upon the unifying effect of principle, and method, which finds its way through manifold phenomena, become the modern manner of recognizing principles and facts as parts of science.

The third possible arrangement of the parts of science to make a pattern of end and means, and consequently a method, has characterized the last two hundred years. The establishment of fact is the end. Since the facts are composed of data, both sensuous and operational, and since the facts obviously fit together in many patterns, principles and hypotheses lend their aid by turning themselves into rules of combination. At first there is a concerted attempt to break down the data into their atomic elements, and this is done so thoroughly and exhaustively by the British philosophers Locke, Berkeley, and Hume, that the connecting forms, abstractions and relations, seem to be dissolved by argument, a beautifully executed suicide of reason. But the reader who is philosophically aware will enjoy the sleight-of-hand, both as a display of wit and as a highly skilled transformation of reason into apparently arbitrary rules for the artistic making of scientific objects out of the finely ground dust of experience. The witty countermagic of Kant's *Critiques* consists in calling the tricks of the British magicians, and showing that what they call habits of combination and interpretation are spontaneous, unavoidable, and necessary functions of the human mind responding with regularity and validity to the abstractions and relations without which there would be no human apprehension of the world.

One might expect that this Kantian counterreformation would reestablish principles and hypotheses in their traditional prior position, and to some extent this has been the result, enough at least to keep a critical controversy in method alive. But Kant's critical effort was persistent and incisive enough to limit the free use of pure reason in speculation, and to harness it to the engines of empirical investigation. For Kant ideas are primarily regulative; that is, they are rules for the operation of the human mind. Some of the rules may be found identical with the course and direction of change in nature, and thus build themselves into natural processes. But even then they are the rules or laws of nature, perhaps even laws governing the operations of the divine mind, as a theologian might see them. In addition, there are merely regulative principles which guide research, but do not constitute things. Principles of this sort govern the art of hypothesis. Hypotheses are the products of creative imagination working in harmony with reason applied in the analytical understanding of the objects under observation.

Of course, the Kantian critique was a necessary piece of dialectic to penetrate the confusion of principle and hypothesis that had resulted from their identification in systems of the world during the Renaissance. Kant showed how the angelic ambition of the systematic mind had led to the aesthetic illusions of the scientific Quixotes and to the ugly logical contradictions of the Cartesian Demon.[3] It was necessary at least

[3] "The aesthetic illusions of the scientific Quixotes" refers to the assumption by Newton and his followers that space and time are realities independent of the human observer and his "intuition." The "Cartesian Demon" refers to the deceiving God that Descartes assumes for the sake of argument at the end of Meditation I of his *Meditations on First Philosophy*.

temporarily to cut the speculative thread of the uncritical hypothesis, and to bring it back to earth. He brought it back to the human world where experiment tied it to observed data in the laboratory.

Many of the sciences, as we now know them, had been called the philosophy of nature during the Renaissance. From the time of Kant they have been called physics, chemistry, biology, and so forth, and most scientific treatises open with a declaration of independence from philosophy rather than with a dialectical discussion of first principles. Newton paid tribute to the old style in expounding the "mathematical principles of natural philosophy," but he claimed he made no hypotheses and said he preferred to see geometry as the more perfect part of practical mechanics. So Lavoisier and Dalton in chemistry, Faraday and Clerk Maxwell in electricity, Fourier in the theory of heat, Darwin in biology, Sydenham and Bernard in medicine never let principle and hypothesis outrun fact.

Mathematics, as one might suspect, alone kept the courage of its speculation, and has only recently submitted to a positivistic style and method, having seen with some surprise its wildest speculations caught and verified in complicated patterns of fact. At first, the non-Euclidean and the four-, five-, and n-dimensional geometries, as well as the new algebras with their imaginary and irrational numbers, looked like a new penetration of the realm of heavenly hypotheses. Although it is still possible for a Renaissance mind to see them that way, such imaginings are now for the most part judged vain and empty, since their subtlety and refinements can be saved and put to work in the applied formulae by which rich findings of fact are counted and measured. Perhaps the present terminus of this reduction of principle and hypothesis to rules of operation is to be recognized in the giant superhuman calculating machines that obediently sort, memorize, and rearrange the mountains of data that the new method has collected. Principles and hypotheses first became ideas in the human mind, then verbal and mathematical symbols and fictions, and now the servile devices in thinking machines.

This subordination and exploitation of the more speculative parts of science in the search for data is called positivism in Europe, pragmatism in America, and dialectical materialism in Russia. It marks the temporary eclipse of speculative truth as the end of science and reflects the utilities of the factory and the engineer in its stead.

This attempt to see the whole history of science in one grand view through the pinhole or lens of method may suggest that there is a deep-lying dialectic in the historic process. If there is such a process, the one suggested here by permutation and combination of the parts of science is probably not it. Neither does that of Auguste Comte, following the earlier one of Vico, in which religion, metaphysics, and science follow each other in ponderous order, seem plausible at present. All such stories are tangled and held in suspension by the myth of the Golden Age and the myth of progress. Each of the three methods determined by the ordering of parts can be found operating side by side in each period of our recorded history, and none is exclusively used in any stage of the life history of a special science. Nevertheless, there is a utility for such schematic patterns for the reader of great books, particularly those which report authentically what the scientist of some particular time thought he was doing.

Principles: The Aim of Method

In accounts of method by great authors, there is always something that achieves clear formulation, and something that escapes it; there is something that the author is saying and something that he is doing and not stating. This is most true and most disturbing in the case of Aristotle, who is often named as the father of the sciences. In him the labor of formulation is tireless and elaborate. As Plato had been before him, Aristotle is concerned with the art of hypothesis, but he is even more concerned with the discernment and formulation of principle. In his writings he does not seem concerned with the ways and means of collecting stuff, but there is abundant evidence that a great labor of collection had been done before he wrote.

Aristotle sees a science as a structure in which there are first principles, last elements, intermediate causes which connect the principles and the elements. No science that he had or made is complete in all these respects, as far as we know, but these items of structure are aimed at and exemplified. As previously mentioned, these items correspond roughly to the parts of science that have always figured in method.

Aristotle's greatest achievement was the adumbration of principles. The labor appears to be chiefly dialectical, the trial assumption of a proposition, the definition and analysis of its terms, and the exploration of its consequences. The result seems to come about by the refutation of the proposition by a process like the *reductio ad absurdum* of mathematics, or the refutation of its opposite or negative, or by some combination of these two dialectical devices in serial application. This seems to be the nerve of his arguments, but there are many variations, as one can see in the argument concerning the principle of contradiction in the *Metaphysics,* the argument concerning change in the *Physics,* and perhaps most brilliantly in the argument concerning Zeno's paradoxes, again in the *Physics*.[4] These are the discursive and explicit formulations, but by noting several hints in other places, notably in the discussion of the intellectual virtues in the *Ethics,* it is clear that the grasp of the principle, although helped by dialectical argument, is not accomplished by it, but by a direct intuitive act of the intellect. This is mysterious until the reader under the influence of the argument seems to do this act for himself. The act seems to be induced by the argument, is not the direct result of it as a logical consequence, but comes as a light in the mind, casting light on many other things. Of course, something like this must be the case, since principles are first principles, not derived or derivable from other sources. This experience on the part of the reader is often misunderstood; he has come under the master's spell and his skeptical powers have been put to sleep; apparent insight must be a matter of credulity. Aristotle is a medicine man. But this is a mistake, as many witnesses will testify in good faith. The test is: once having the insight, try with all your powers to think otherwise. If one has actually seen the principle, then the dialectical argument will show the impossibility of thinking otherwise. If one has not seen the principle, the dialectical argument will be weak and unconvincing, or perhaps worse, arbitrarily dogmatic.

Intuitive Induction

Hypotheses are arrived at by a process which Aristotle called intuitive induction. There are of course many processes in science that are called induction, methods for discovering in or abstracting from the manifold data those generalizations, correlations, and uniformities which make the familiar fabric of scientific discourse. There is much evidence that Aristotle had collected, classified, and correlated many things. He even talks about induction by exhaustive enumeration of cases, about observation of similarities and differences, about temporal and spatial associations of phenomena, and indeed about crucial experiments, but for him these are not the decisive factors in the process. They are either random gropings before judgment is made or they are the mopping-up operations after the strategic stroke of mind. Hypotheses are opinions that can be treated dialectically for purposes of refutation and further search. He compares them to the wandering and dissolving parts of a

[4] *Metaphysics* Gamma, 3–6; *Physics* I.2–3, VI.8–9.

retreating army, and asks how it is that such an army rallies. His answer is that one soldier, maybe an officer, or maybe a private, turns around and makes a stand. The others seeing him, imitate his action and remake the battle line on him as a pivot.[5] So it is with primary induction; an observation identifies itself with a hypothesis, and the other data line up in an order dictated by the hypothesis, or they disappear. Intuitive induction is the apprehension of the hypothesis in the datum. This is good description of original discovery in science, as every scientist knows. The rules for induction all depend upon this original apprehension and they help to make it effective in building the chance stuff into patterns of expanding intelligibility.

The so-called rules of induction, as they have been compiled and revised from the time of Francis Bacon to the present positivists and pragmatists, all deal with secondary inductions. Given a guess or a hunch about some similarity, some uniformity, or some repeatable pattern of relations, the rules that flow from the guess provide a kind of sieve or crosshatch for saving and ordering the appearances, but they reveal few clues about their sources. Like the dialectical processes that surround and support the discovery of a principle, they seem to set the conditions and support the search for the center or pivot on which they swing. After John Stuart Mill had reviewed and patched the rules that Hume had laid down, he announced that the great unsolved problem of scientific method was the explanation of the fact that the great generalizations of science had been verified by single cases. For many reasons, verification is a very modern way of viewing the problem of induction, but Mill was backing into the center of the process. He probably missed its solution because he was looking for rules where none can exist. Primary induction is an original act, of the kind that we moderns call creative. It may be the inevitable way for an angelic mind to operate, but for humans it is a gift; as in the dialectical discovery of principles, intuitive induction is something that the human mind does and recognizes after the fact.

The Greek word for induction had another usage, as many technical scientific words had.[6] It refers to the introduction of a case into a law court. An event has happened, and there is a conflict of opinion about it. Its reasonable treatment by the human community involves the formal procedure of a court of law. It is the business of the advocate to formulate the narrative of the event in such a way that it will fall under a rule of law and determine the relevance of the evidence to be offered in support of a plea and a verdict. The advocate has the task of finding

[5] See *Posterior Analytics* II.19, especially 100a9–14.
[6] The Greek word referred to here is *epagogē*.

the approach and making the proper entry. Although there may be precedents which will indicate the approach, there will always be some need to construct the case. This is true whether there is the intention of merely winning the case, or whether justice and truth are to be served in addition. Apparently, in Athenian law courts many of the cases were constructed in imitation of patterns in Greek tragedy; the craft of the playwright in making plot define character and action played its part as it still does. So the hypothesis must be invoked to save the appearance in science. The question for the investigator is what role the phenomena play in a hypothetical plot. A fact can enter a science only in some role; there must be a rite of induction. For a Greek there was a general question in both law and science: How does a case become a cause? We can get the feel of this question by asking how an appearance becomes a fact.

One of the simplest and most elementary instances of primary induction occurred in Aristarchus' determination of the relative distances between the sun, the moon, and the earth.[7] Among all the appearances of these bodies, he noted one as crucial for his scientific purpose; that is, when both the sun and the moon are in the sky simultaneously, and the moon is at half-phase. Among many thoughts that might and probably did occur to him, he noted one that gave these appearances a role: the angle between the line from the sun to the moon and the line from the moon to his eye must be a right angle. This direct identification of a pattern of light and shadow with a geometrical relation was an original intuitive induction. It led immediately to the construction and identification of the three bodies as the apexes of a right triangle. With this triangle as a frame of reference it was easy to conclude by Euclidean inferences from this and a few other observations that the distance from the earth to the sun is twenty times the distance from the earth to the moon. Since Aristarchus' measurement of the other angles in the triangle was in error, his result is considerably in error, but the hypothesis of the right triangle is valid and can be used today.[8]

It is possible to see in this induction something that looks like steps in sequence: first a collection of observations of heavenly bodies in

[7] A translation of Aristarchus's treatise is given in Morris R. Cohen and I. E. Drabkin, *A Source Book in Greek Science* (Cambridge, Mass.: Harvard University Press, 1958), pp. 109–13.

[8] Aristarchus's result is too small by a factor of about 20. The chief difficulty in the method is the determination of the exact time of "dichotomy" or half moon. Wendelinus and Riccioli were still using the method in the seventeenth century and obtaining results that were off by a factor of 2 or 3. The more recent determinations of the sun's distance have been based on phenomena other than dichotomies, for instance the transits of Venus.

relation to the earth, then attention to an apparently unusual coincidence, the moon and the sun in the sky at once, then the sorting of geometrical figures with a leaning toward the well-studied case of the right triangle, and finally, the selection of the half-moon as a point of identification for the two systems, celestial bodies and geometrical figures. But these steps yield no rules, no frame of reference, no inevitable generalization. Although all the materials are familiar, the original birth of a powerful generality is novel and unpredictable. It is as simple as the childlike identification of the constellations, but it has a scientific power which they only suggest.

But this induction is not a unique case. The Greeks both before and after Aristotle sparkle with insights like this, not only in astronomy, but in physics, biology, and medicine. It is not an exaggeration to say that the rest of our scientific tradition has been the laborious application of the rules of secondary induction to the world which they penetrated and marked out for development by primary intuitive induction. Perhaps the most characteristic and fateful operation of this intuitive power in the Greeks occurred in mathematics. Most of our modern mathematics is a monument to this heritage, but the style of the mediums in which it is reported to us would tend to distort our vision of it. Euclid's *Elements* and the uses we have made of it in education seem to be evidence that deductive proof or demonstration is the essential nerve of mathematics, and that the discovery of this technique is the characteristic contribution of the Greeks. The fact seems to be that Euclid did a typical encyclopedic job of sorting and compiling other men's works. His sources would show a great mass of independent discovery and fragmentary development, based for the most part upon the insights gained from trial-and-error rumination and contemplation, as well as upon the solution of problems in other fields of inquiry. Mathematical proof is a powerful systematic device for sorting and certifying mathematical knowledge, but it must be clear that it is itself an artificial fabric made of insights that have been achieved in many other ways. Demonstrations depend upon direct apprehension or they are a mechanical fraud. They are security devices for certifying and keeping the real mathematical knowledge in mind and at hand.

It may not be too much to claim that the whole of Greek mathematics is the collective result of thousands of intuitive inductions. There are theories of its origin which would deny this, and attribute its existence to divine revelation or to the invention of symbolic and pragmatic tricks. The participation in the mathematical enterprise that the reader of Euclid, Apollonius, and Archimedes achieves gives one the habit of seeking and recognizing the illuminating insight into familiar things rather than the stealing of heavenly fire or the stealthy tricks of the practical man who is cooking up solutions to problems.

If this is so, it is not an accident that mathematics is so frequent an ingredient in hypotheses. The right triangle that Aristarchus anchored in the half-moon, the numbers that identify themselves with shapes and sizes, and above all the ratios that express the intervals between musical tones, and the regular solids that give form to the elements—all of these objects which at some stage of induction float freely in the imagination have affinities for facts that seek to become characters in the drama of nature.

But hypotheses are not exclusively mathematical. There is a reading of Plato that would all but conclude that the ideal destiny of a hypothesis is to become a number; there are schools of Platonism that would turn even principles into numbers. It is true that there is a class of ideas intermediate between images and principles which he called mathematicals. He often also calls them hypotheses. But for Plato mathematics had almost its literal Greek meaning: mathematics was learning, both for pupil and for wise man. But learning by hypothesis and mathematics is connected with Plato's concern with a possible medium by which ideas and perceptions can get connected to make science. The mathematicals are the beginning of learning because, as numbers and figures, they count and arrange things so that they can reflect ideas; or in reverse, the mathematicals can translate the unity and identity of ideas into an order and relatedness so that they can exist in many changing things. The mathematicals provide the orders and patterns in which ideas and things can co-exist without mutual contradiction and annihilation. It is because of its penetration into both ideas and things that the intuitive induction so often uses mathematics to give the mind a glimpse of the way to valid generalization.

The controversy between Plato and Aristotle concerning the mathematical forms, although it seems to be primarily concerned with the ultimate metaphysical reality of numbers, arose from a discussion within the Academy concerning the role of mathematics in science. It reflects work in mathematics, astronomy, and physics, but it also reflects the birth pangs of biology, ethics, and politics. There seems to be an agreement that mathematics has a place in all these sciences, but the degree to which mathematical methods and analogies are used in each is a matter of some argument. A Platonist in the Academy extrapolating from his successes with the irrational numbers and the harmony of the spheres seems to have made the familiar claim that when the other sciences come to maturity, they will be found to have become mathematical. The Aristotelian answer seems to have been that although all things may be considered mathematically, their mathematical treatment is merely preparatory to their proper treatment, which will be derived from categories other than the quantitative and relational. The argument goes so far that even figures and numbers are treated as qualities. When the smoke of controversy clears away, it appears that the role of mathematics has been limited; mathematical objects are certainly not substances, although their clarity partially justifies their consideration as separate independent things. The limitation is shown by the equal though different clarity that genera and species in other categories can achieve in the scientific process. As there are figures in geometry, so there are figures of the syllogism, and as there are continued proportions involving the tight articulation of ratios, so there are sorites which show the internal articulation of forms. It is as if the Aristotelians had been challenged by the power of mathematical insight to show the equal and more comprehensive power of logical insight. As mathematics, in the form of a right triangle, provides the framework within which the half-moon indicates the relative distances of the sun, moon, and earth, so does a desire show the capacity for virtue and the distance a man is from happiness. Powers, habits, and acts make a nonmathematical right triangle, and allow an intuitive induction in the science of ethics.

Plato's dialogues are full of arguments and counterarguments that are based on intuitive inductions of many varieties, but he did not give them the name. Aristotle, bent on laying the foundations for the various sciences, collected and systematized them. He saw the intuitive induction as the way facts have of learning through mathematics or logic how to become truths; he formulated as method what Plato had abundantly and variously practiced.

Surface and Artifice

NOWHERE is the discrepancy greater between the ancient mind and modern mind than in their respective views of the empirical practice of scientific investigation. The ancient medical arts of diagnosis and prognosis, in which the close observation, recording, and interpretation of symptoms achieve elegant precision, come the closest to the modern notions. In physics and astronomy there is a great deal of implicit know-how which never reaches concrete verbal expression, but is discernible only by sympathetic guessing on the part of the modern technician who has the feel of the laboratory. Aristotle is clear and explicit only concerning the role of the senses in these processes. There seems never to be a doubt in his mind that whatever we sense in a natural object exists as a singular concrete quality in that object. Nevertheless, the sense datum, as it exists in our knowledge, is a sensible species, a sensed universal which immediately subsumes itself and is comprehended in a hierarchy of genera and species. Thus when we sense the yellow in a lemon, we immediately know that yellow is a color, and that color is a quality that can exist only in a surface. The basic scientific observation is therefore already set in an intelligible order and is therefore capable of entering into a pattern of hypotheses and principles. This seems to ignore the primary difficulty and to take the mystery out of intuitive induction, but there are many warnings that it is a mark of special genius to spot the strategic fact and see its connection with the proper hypothesis. Experience and observation provide too many suggestions of hypotheses; it takes a Theseus to follow the inductive thread through the labyrinth.

The less explicit and less clear account that Aristotle gives of observation deals with the artificial imitation of nature that we call experimentation. His work is full of evidence that he is aware of the gadgets that the Greeks used—levers, wheels, and pulleys, as well as measuring devices and manual tools—and he is constantly referring to the skills and rules of the crafts. He is perfectly aware that these are imitating natural processes of change. But he does not seem to be fully aware, as many of his faithful followers are, that all these things are the extension of the senses and the memory, and that they are the operations that exemplify hypotheses and principles. Galen, for instance, repeats again and again, like a proverb, the principle that art imitates nature, and that therefore nature, especially in the living organism, is a superior artist. Galen of course is here quoting Plato much more than Aristotle, but all of Aristotle's theory of nature is founded on this analogy. At any rate, it is quite clear that the human imitation of nature argues in the artist a basic knowledge of nature, and that much theory of nature is due to an abstraction from this knowledge. It perhaps takes the kind of concentrated and organized attention that moderns have given to experimentation in the factory and in the laboratory to bring the full realization of the importance of the processes of making in the scientific enterprise. It is seldom given the dignity of explicit formulation, but it is honored universally in what the scientist does, this principle that if you can make it, you can understand it.

Of course the principle, thus roughly stated, does not mean that making and knowing are the same thing; it does not mean that the artisan in a man and the scientist in a man are the same thing. But it probably does mean that there is no science without artisanship. The relationship between the art and the science is indicated somewhat in the two ancient tests for man's rationality: he makes things with tools and he talks. Both of these are arts, kinds of making or *poiesis,* but they are quite different, part of the difference being shown in the power the talk has to reflect the making. But this reflection is a kind of knowing, not only the knowledge of the making but also a knowledge of the materials, the forms, the instruments, and the ends of the making. It is out of this knowledge that Aristotle refined his theory of the four causes, the material, the formal, the efficient, and the final. Their hypothetical imputation to natural processes is the main strategy of scientific investigation. It is out of them that hypotheses, which are the specification of them to fit the case, get their schematic power to arrange and order the data. As the skills of operation sharpen the senses, so the hypothetical causes determine the relevance and the roles of the facts.

Logic and Metaphysics

So FAR, this account of ancient scientific method, as the reader has probably unpleasantly recognized, has followed a devious route, piecing together fragments of information and dialectical discussion in the writings of Plato and Aristotle with scattered references to the more specialized scientific work in Aristarchus, Galen, Archimedes, Euclid and others not named. The resulting pattern, as here presented, is in all likelihood one that would not be accepted by any one of the original authors. This is partly because none of them would have consented to lay down the rules of his method, or have considered his writing about the scientific enterprise as proper material for the formulation of method. Their talk about the great speculative enterprise is useful to us, not because they were intending it to be used as we use it, but because they were good workmen immersed in the work and at the same time conceived it so clearly that their logical and metaphysical shop-talk is as good as our methodology. There are a few modern scientists whose brief, incisive shop-talk is authentic in this way, far better than the mountains of methodology that we have accumulated. The fact is that we have chosen to do the inescapable metaphysics and logic that goes with our science in the medium of methodology, and we have to see the job through.

We pay for our indifference to logic and metaphysics in many ways. Our methodology becomes labored and fussy about minutiae; we find the history of science full of myth and perversion, and as a concomitant of this, misread the past. Many able and original minds have misread Aristotle in the last three hundred years, particularly the five treatises that make up the *Organon,* the description of the instruments by which science is done. It has been assumed that these treatises were Aristotle's methodology. A reading of these treatises in the context of the scientific enterprise as it was going on makes them seem less and more than methodology. They are less than methodology because they give very little account of the techniques and skills of the scientist; all these are taken for granted. They are more than methodology in that they are highly formal sciences, pitched on a very high level of abstraction and maintained on this level only by eliminating and ignoring the concrete and operational content. Their content, if any, is scientific discourse, a discourse which logically comprehends all manner of content but achieves its end of formal universality only by suppressing its applications. The attempt to expound and criticize them as methodology is bound to produce illusion, which is easily refuted but never cured in modern criticism.

These treatises deal with the clarification and ordering of terms, the analysis and classification of propositions, the structures and figures of syllogisms, and the larger structures that can be built from these parts under the general name of sorites. Many followers of Aristotle as well as his opponents, by taking this material as methodology, have drawn a swift conclusion from it that science for Aristotle was summed up as deductive demonstration. Whatever does not achieve this discursive form is not science, or if it only partially achieves it, it is probable inference. This bad reading, both by the defenders and critics of Aristotle, not only falsifies the past, but also impoverishes the mind that tries to deal with any scientific problem of the present. It hides the problem which the ancients did so much to state.

The doctrines of formal logic would never have been stated had it not been for the actual achievements of the Greeks in the various sciences. These achievements were at the least the discovery of a set of principles in the context of a well-developed dialectical discipline, the anchoring of hypotheses in strategically selected fact, and the orderly observation of nature by hand and by eye under the guidance of both principle and hypothesis. Each of these processes has its own validity and certification. It was the function of formal logic to bring these different knowledges together in one view without sacrificing the integrity and validity of each part, and in such a way that the resulting science would be a knowledge of reality, not merely a construct of sense and imagination.

The task was set for formal logic in its most general and comprehensive scope by Plato in the *Sophist.* This is a late dialogue written after most of the dialectical pursuit of ideas had run its first great course, and it was time to return to the home base to get an ordered view. The first part of the dialogue indicates a perspective. Although the playful wit of Socrates had exploited many a trope of logic and even of sophistry, Plato saw the possibility of reducing all these versatile tools to one simple knife, the knife that uses the cutting edge of the negative to divide any universe of discourse into two parts, horse and not-horse, straight and curved. Although this device runs into paradoxes with some extremely general ideas—for instance in the tight dialectical patterns that trace the connections between being and nonbeing, the one and the many—the negative cut reveals the underlying structures that have to be recognized if thought is to be faithful to the distinctions in things. For instance, the distinctions of principles, hypotheses, and facts are almost immediately revealed by the cutting of any universe of discourse. The principle is the most general concept or universal. It does not all by itself divide the universe into two parts; there must be another quasi principle having a partial identity with the king-concept but relatively limited in its comprehension so that its negative shares in the common property. Suppose 'mammal' is the defining concept for the field; then 'horse' and 'not-horse' divide it between them. But horse is a somewhat arbitrary choice, it might be 'mouse' or 'man'. One might even introduce 'imaginary animal' and 'real animal' as the cut. Such proposals are hypothetical, but like many hypotheses bring definiteness and the possibility of inference into the field. By repeated application of hypotheses and sub-hypotheses there appears to be generated a search for the last distinction within which a concrete individual or lowest species of individuals can be caught, identified, and named. Plato discovered that there can be any number of such model sciences built within any community of ideas, and that these little sciences will overlap and cross-refer themselves. Like an explorer he drew a map of the area in terms of the basic meridians and parallels with the poles around which they organize themselves. He noted that some ideas, such as 'being', 'one', and 'same', coalesce dialectically; they identify themselves with each other and all other ideas. Ideas with less generality sometimes overlap and subsume themselves, such as 'mammal' and 'horse'. Still other ideas stand apart radically like unique individuals. Some of this classification seems to be relative, some of it seems to be absolute. It is with this rough but comprehensive outline that Plato gives the assignment to Aristotle to order the field of thought. Aristotle's response is found in the five treatises of the *Organon.*

Aristotle's analysis takes the negatives and oppositions of Plato's dichotomous knife for granted, makes the cutting of the universes systematic and precise, and results in a doctrine of univocal or unambiguous terms, a machinery of predication that classifies propositions, a scheme of combining propositions into arguments and demonstrations, and a matrix of terms that reaches from highest generality to lowest specification and supplies all discourse with patterns of implication. This is an amazing performance; seldom have two men asked and answered questions so well.

It is not surprising that this imposing achievement laid the foundations and indicated the future construction of the sciences for the next two thousand years. Neither is it surprising that many of the later workers did not see the major outlines, and fell to fatal applications of their misunderstandings of it. There have been many misunderstandings on both sides of the quarrels that have ensued. But perhaps the worst precipitate of these quarrels has been the predominance of the formalistic deductive logic. When the vital insights that underlay that part of the ancient work have disappeared, the logical machinery has been multiplied and speeded up. By wooden minds formal logic can literally be built into machines which for science can emulate the feats of prayer wheels in religion, or calculating machines in engineering schools. Whenever this happens, the intellectual light of science, which never completely goes out, seems to come from the top of a hierarchy of concepts. Certainty is conferred upon contingent and doubtful matters by deducing them from first principles. One very late formulation of the ancient method in the fifteenth century runs as follows: Science starts with doubtful fact, looks for certain first principles, and then by syllogism demonstrates the certainty of the fact from the principle as first and major premise. This seems to ignore intuitive induction, and puts the first principles beyond the reach of dialectical and experimental revision. Much of our misunderstanding of scholasticism is due to the wooden orthodoxy of its second-rate adherents and practitioners.

It must be said that there were continual recoveries from this dry rot of thought, and when these happened there were not only revisions and new discoveries within the framework, but there was always the understanding that high abstractions were merely the highest reach of a deep unification of thought that demanded and accepted the evidence and insights of the senses and the intellectual imagination.

There is nothing merely formalistic in Aristotle's own work. The mountains of material that have the marks of his own observation upon them, the measured hypothetical construction in terms of which the material is deployed, the double and triple dialectical approach to principle, and the almost miraculous eye for the strategic generalization need and justify the formal structures of logic for their containment. And the logic plays two roles; one suggested above is like mathematics, the extraction and formalization of what he called the essences and the proper accidents of things so that their crucial and necessary connections could be seen as causes.

Modern mathematics has a greater versatility and power for this role than the cumbersome terminology of Aristotelian logic, but it gets this increment of power by dumping some of the significance of the data it correlates. Aristotle's logic, put to work in physics, dealt with six different kinds of change, substantial change or coming-into-being, passing-out-of-being, growth, decay, alteration or qualitative change, and locomotion or change of place. Ancient mathematics together with a kind of organon of analogies could deal with all these changes and bring measurement of many kinds to many things. Modern mathematics reduces all these motions or changes to local motion. We pay for this gain in power by a loss in our ability to see the unity in the natural and human world. Our significant world splits into the two parts of a bifurcated nature, as A. N. Whitehead has called it. The split happened for us in the Renaissance when Greek mathematics was revived through Arabic algebra, to which the next discussion of method will turn.

Aristotelian logic kept mathematics in suspension and used it as a medium of knowledge, and it maintained those modes of signification by which a responsibility to reality and being is assured. A term in Aristotelian logic is always raising a question in the mind about the reality of its object. The copula *is* in a proposition is always reminding the thinker of the great subject of all propositions, being. Aristotelian logic is therefore always indicating and accepting a metaphysical responsibility. There is here no ontological argument to demonstrate that if you think, therefore you or anything else must exist, but rather a persistent demand that if you think, you had better be thinking about something that is. The demand is based on a quasi-intuitive induction: to think is to think about being. This is the ground on which the scientist stands when he inquires; it is, whether he likes it or not, his first and last commitment, no matter what hypothetical journeys he may make in his mind. This is the speculative nerve that keeps science faithful to its polar star, the truth. We pay tribute to this faith in our modern confusion by the gnawing and sometimes proud awareness that we are making progress without knowing either our direction or our destination.

The price the ancients paid for keeping their faith was of two kinds. Their data were perhaps too much restricted to common-sense observations and the factual by-products of work with manual tools. On the other hand, astronomy was perhaps too fascinating to them because it dazzled them with visions and intimations of an infinite speculative realm. The other price they paid was an almost incredible inhibition in the kind of imagination and invention that has led to modern industrial development. One of the most teasing of the unsolved historical questions concerning the Greeks is why they did not bring off the industrial revolution which the revival of their science produced or helped to produce in the Renaissance. Part of the answer must be that science for them was the art leading to the good of the intellect, the truth, not a handmaiden of the utilities. The data of the senses, the storehouse of the memory, the poetry of the imagination, all these were bent to the speculative enterprise. It is not enough to attribute this bias to the privileged class of freemen who kept their freedom by exploiting slave labor, since the slaves contributed heavily to the arts and sciences. In terms of the scientific enterprise, it seems more plausible to give credit and blame to the intrinsic importance and vital drive of intellectual curiosity, and to attribute to the economy of that interest the upward drive from sense and hypothesis to principle. For the Greeks the intellect belonged to the natural world, and it was recognized and trusted when it was demonstrating its affinity with being. Like Plato's philosopher king, the intellect returned from high visions to illuminate the cave of the world. What we take to be their blind search for light resulted in our cave being lighted as well as it is.

The Rebirth of Ancient Science

For a long time the Renaissance has meant in the clichés of history the rebirth of ancient culture among the ruins of the medieval world; it has sometimes seemed to be an almost miraculous case of spontaneous generation, and our understanding of it in this light has thrown a romantic glow over our view of the ancient world. But as historians and other scholars have collected and recollected more and more of the circumstances of the rebirth, it appears that the offspring had a mother and that the principles of genetics were in effect. The more natural conception of the Renaissance can be seen and admired in Dante's *Divine Comedy* where the germ of the ancient world lodges, lives, and grows.

Arguing from effects to causes, the lively genes in the embryo are love and mathematics, and it is not unlikely that they were transmitted from the two prose poems of Plato, the *Symposium* and the *Timaeus*. It is noteworthy that the *Divine Comedy* was delivered in exile and that its two themes were heretical and revolutionary in Greek thought.

Although Dante held them with the help of St. Thomas in an Aristotelian suspension, the child in whom they lived was bound to have a stormy, powerfully creative life.

Dante was first and always a poet, and for a long time he has been read as if he were a modern poet of moonlight and feeling, but like Homer and Virgil and Shakespeare his greatest power is intellectual intuition. His vision is persistently intellectual and he always has the courage of his curiosity. He will see and understand before he imagines and speaks. His sight is almost identical with Plato's where the objects are the same, but he also sees Plato, Aristotle, Euclid, Ptolemy, and others whom Plato could see only in shadows and reflections. Dante very often sees not only the themes of mathematics and the points of morals, but manages to see them together, as in the circles of vice and virtue that define his geography. In his mind all these visions gravitate to their proper places. It is incredible that one man could see and compose so many things, until one notes that it is the Aristotelian framework that stretches itself so that it can comprehend without crushing the independent apprehensions and can focus without smothering the separate fires. One dimension of the framework is added. Where Diotima has Socrates stretch his vision to look upon beauty-in-itself, and Aristotle pushes his logical machinery backward in the series of causes to the unmoved mover which is both the final and efficient cause of all causes, Dante adds an apex to the hierarchy of forms, in which love and knowledge combine in the beatific vision of the single essence of all essences, a view of all things as they are. This is most surely the mystic vision, but Dante is able, as most mystics are not, to mediate and temper this vision to accommodate the things seen and the human seer. He does not sacrifice possible human vision to impossible divine wisdom. He makes the whole, which all method seeks, the principle of the itinerary to it. Human vision, which is a fragment of divine wisdom, views principles, hypotheses, and concrete person and fact in glasses that adjust and readjust their focuses silently; even human blindness itself is an essential episode in the journey.

This is truly the rebirth of the Greeks in a form that would have dazzled their eyes, and yet they could not have failed to recognize their new selves. Dante himself has a hard time keeping most of them in limbo, and in fact, he uses their minds most crucially where he himself finds the ascent steepest and most confusing. This world of Dante's is so full that it must spill over, as God's love is said to do. Either it spills over, or it breaks from internal pressure of being. In certain ways it did both, and it poured out in two great streams of thought—and of course of feeling. These streams are Greek recognizably and certainly.

One of these is the philosophy of love, which stems originally from

Plato's *Symposium* and issues in many treatises on the passions of the soul, reappearing in our time in the writings of Sigmund Freud. The ancient insight is that all the passions in man are forms of love, as all the changes in nature are varieties of natural love. The exposition of this theory is placed in the very middle of the *Divine Comedy,* the seventeenth canto of the *Purgatorio,* so that heaven and hell, as well as nature, can balance on it as a pivot in Archimedean fashion. Everything in existence is on a circle at some remove from the central *summum bonum* which is God. Each existent is a vector pointing away from the object of love, toward it, or tangent to the circle. The proper direction of each vector is toward God, but the variation from this norm is caused by local disturbances among the apparent goods. These perturbed vectors may get stuck by habit, and they will then measure the vices and virtues of men or the variations in perfections of inanimate things. But the pull of the central good is always a component in the pattern that orders the realm, and in man the intellect and will are always the faculties of the truth and goodness of the central good. This doctrine is never wholly forgotten in modern thought, whether it be expressed in Botticelli's illustrations for the *Divine Comedy* or in his *Primavera,* in Schopenhauer's *The World as Will and Idea,* or in the latest novel.[9] It is the fundamental language of the romantic poet as well as of the newest school of psychology. It is perhaps most honored in its belligerent opponents, of which there have always been many from the Greeks until now. Freud avoids explicit acceptance of the doctrine only by going back to the best original in Sophocles, who saw the human predicament with a single honest eye, and from whom Dante himself learned about the role of human blindness in the economy of love.

It is true that the Greek search for the *summum bonum* as the adequate object of love, a search that tired into a melancholy disillusionment in the Roman account of it as voiced in Lucretius, found its speculative satisfaction only in Christian theology. But even in the New Testament, where the doctrine that love is the light of the world is promulgated, the language of the answer and the terms of the search are Greek. χάρις, *caritas,* and charity carry the doctrine of grace and of the theological virtues to St. Thomas where Dante found them reflecting the Courts of Love and the practice of the monasteries. This stream of thought waters and permeates every last corner of Dante's three realms in fountains and cascades of many colors.

The speculative and lyrical height of Dante's poem is, of course, not

[9] For an interpretation of Botticelli's *Primavera* that accords with and fills out Buchanan's suggestion here, see Edgar Wind, *Pagan Mysteries in the Renaissance* (New Haven: Yale University Press, 1958), pp. 100–110.

an isolated phenomenon. If Dante had not been exiled and thus forced to rally his spiritual powers, someone else would have had to scale the Alpine heights of the medieval *summas* and bring the many-faceted visions that went into their construction into a single focus. The labor, study, and contemplation of the monasteries were in the *summas,* and several centuries of disputation in the universities had articulated the intellectual workmanship of their structures. They have often been compared with those monuments of the collective mechanical arts, the cathedrals, and they might be compared with those works of literary art that penetrate and elaborate the sacramental system, the liturgies of the church. Not all of the summas penetrate the mysteries; like many cathedrals and much liturgical invention, the *summas* are sometimes contraptions of scaffolding. Some of them are composed of indestructible thought, and will last forever, but they have no internal fire and no transparency. But others glow and radiate light like the Biblical burning bush.

Literature is often examined by the critics to see if it expresses the spirit or genius of the time. The critics' conclusions are often, as in our time, mysterious and romantic with regard to the mediums of responsibility and communication which such expression involves. The literary artist moves over the face of his world and divines the hidden streams and currents that he reports. In this kind of literary criticism there is a nostalgic imitation of medieval craftsmanship. When Thomas Aquinas quotes an authority, he is establishing a definite relation, often of logical opposition, to a source of his thought. He is using a tool of thought fashioned by a predecessor to push his own mind deeper into the question, a question that has been asked many times before by many different minds and elaborated into many articulated points. The responsibility of the writer of a *summa* to communicate with other minds as well as with the subject matter is thorough and explicit. The spirit of the time has a body and a style that is vibrant and communicative throughout. Dante belonged to such a community and his poetry is therefore a penetrative summary of *summas,* as he himself frequently acknowledges specifically.

The *Summa Theologica* of St. Thomas is one of the burning bushes, and the central fire that it radiates arises from a single insight, an insight into the nature of the human intellect. As a young man he had written his contribution to an Easter disputation as partial fulfilment of the requirements for the degree of master of arts at the University of Paris. It was entitled *De Ente et Essentia,* a theme that went directly to the root of the intellectual problem of the time, the place and function of universals. Ten centuries of church councils, of political controversy, and of scientific argument with some recurrent periods of silence and solitary contemplation had brought the battle of the liberal arts back to the basic question of European thought, what to do with ideas. It took a freshly disciplined, undoubting Thomas to grasp the question and reformulate it. And that is what he did before all the masters and doctors of the university.

The essay *On Being and Essence* probably uses the smallest number of words to convey the greatest intellectual burden of any great book of the Western world. It is a wonder of compression and comprehension. Where other men had written *summas* to defend positions, and had picked other men's minds to prove their points, the young Thomas uses other men's words to complete and fulfill their thoughts and to make thought in general effective. He does this by detecting and exhibiting the human instrument of thought, the intellect, and by tracing out and defining its powers, habits, and acts. As the intellectual operations are the subject matter of the essay, so the writing itself is an eloquent exhibition of the intellect working in discourse. It took Thomas the rest of his life to formulate the detailed implications of his thesis in the *Summa Theologica* and many smaller complementary works, but the central flame from which these lights are kindled burns in the few pages of this first formulation.

Part of the fresh power that goes into this reformulation of the ancient question with which Plato and Aristotle wrestled comes from the Christian doctrine of grace. This doctrine begins with a recognition and acceptance of the sickness, distortion, and frustration which the human mind suffered in Adam's fall from the ideal of intellectual efficiency in the state of nature. The vagaries and failures in discursive thought are evidence enough of a state of sin. Through the Incarnation and the consequent scheme of salvation God makes a free gift of grace through which the mind not only can repair itself, but can also enhance its original powers. New capacities and appetites appear in the state of grace as if by a new divine act of creation, and with exercise they develop into the so-called theological virtues of faith, hope, and charity. Augustine had seen and identified these new virtues; he had even recognized them as parts of the virtue of wisdom as glimpsed by the Greeks, but it was left for Thomas to establish the seat of these powers and virtues in the intellect. For a modern they are the gifts of genius or the result of emotional ferment; for the Greeks they were signs of heroic birth; but for Thomas, whatever their divine source, they were proper extensions of the intellectual faculty. A man believes, hopes, and loves in order that he may understand.

There is a dominant tendency in contemporary modern thought to put faith, hope, and charity beyond "the bounds of pure reason," even to see them as opposing and subverting the right rules of reasoning.

Faith is connected with the superstitious belief in miracles and myths; hope is associated with funerals and a life after death; and charity is a social device for dealing with poverty. The theological virtues, or sentiments, as they might better be called, are the skills devoted to that higher guessing game called the philosophy of life. For Thomas Aquinas they were grateful and graceful increments of the power of thought to enable a man to deal wisely with the ultimate questions. They were the means for glimpsing and possibly grasping the highest ends of both speculative and practical reason. Faith and hope stretch the mind to the true and the good, and they remain as a memory and promise of success after each necessary failure in the life of reason. They are the substance of our confidence in the intellectual enterprise which is so precariously pursued through ignorance and despair. Ultimately, the awareness of an end is more important to a human enterprise than the assurance of the means for its accomplishment.

"And now abideth faith, hope, and charity, but the greatest of these is charity." All of the writings of the Angelic Doctor, as St. Thomas was called, can be understood as the exegesis of this text. All of the moral virtues depending on justice, prudence, courage, and temperance; all of the intellectual virtues, art, science, and wisdom; and the theological virtues of faith and hope die and are turned into vices without charity. With charity all human capacities live and stretch toward their proper ends, energies are released, and the fullness of being is revealed. And all this has a special bearing on the intellect in which the capacity and virtue of charity resides. With charity, the intellect, the eye of the mind, seeks, finds, and loves its proper objects.

It is this doctrine that gives structure and power to Dante's poem, which uses the analytic powers of the *summas* and converts them into a single vision of the universe as a system of love.

Numbers and Figures

THE other stream of thought[10] must have arisen in unrecorded history, its sources dimly but clearly refracted in the Babylonian and Egyptian astronomy. It is the science of numbers, figures, and ratios. It enters the plain of history as rivulets of folklore accompanied by as much mystery, magic, and myth as ever was devoted to the rites of love. All of

[10] The first stream, we may remind the reader, is the philosophy of love discussed in the preceding section.

this is collected and dramatized by the Greeks in the semilegendary history of the Pythagoreans, who founded a secret brotherhood, a school, and a polity in Southern Italy. Their leader, Pythagoras, though not a full-statured hero, has some of the traits of a Hercules or an Asclepias. The society seems to have been an early institute of research in mathematics, the school taught a discipline, partly medical and partly moral, and the polity is said to have suggested certain of the elements in Plato's *Republic*. Fame imputes to the Pythagoreans two major mathematical discoveries, the Pythagorean theorem concerning the squares on the sides of the right triangle,[11] and the isolation of the incommensurable numbers and magnitudes. They also discovered the simple whole-number ratios that control the harmonies and melodies of musical tones; and in another field, logic, they made the catalogue of logical oppositions that later led to the search for the great categories of thought. Their doctrines were apparently capable of further generalization, which issued in the metaphysical doctrine that all things are composed of numbers. Their exoteric teachings were delivered in verse, and this apothegm, like many others of the time, became not doctrine, but rather an oracular question, puzzling and inciting the native curiosity of the Greeks to wonder and search for its meaning. It is perhaps the original of all the good tricks for teaching mathematics.

The Pythagoreans were of course no more primitive and naïve than Homer, their counterpart in poetry. There is irony, subtlety, and a powerful disciplined search for the truth in all this. It turned a lively bag of tricks into a never-ending tradition of thought. Plato's tribute to the legend and the tradition, which seemed old and awesome in his time, is in the quaintest and most imaginative dialogue he wrote, the *Timaeus*. In it the aged and revered Timaeus, a visitor to Athens from Locris in Southern Italy, tells the story of how the mathematical architect of the universe made the things that are in it. He first constructed a completely self-sufficient living animal in the shape of a perfect sphere. The soul or vital principle of this celestial animal was a pair of great circle zones, the equator and the ecliptic, which he called the circles of the Same and the Different. On the circle of the Different he placed a numerical scale representing a continued proportion composed of the great ratios, the arithmetic, the geometric, and the harmonic. The intervals represented the musical intervals, and were capable of further subdivision in terms of other possible ratios, the table for which can be found in Nicomachus's *Introduction to Arithmetic*. Then the stars and

[11] On the question of the ascription of this theorem to Pythagoras, see Thomas Heath, *A History of Greek Mathematics* (Oxford: Clarendon Press, 1960 reprint), I, 144–49.

planets were made as intelligences, and the lower rational animals including man, souls with progressively decreasing intellectual powers. Finally, the material elements were constructed from atoms in the form of right triangles, the hypotenuses of which were the square root of two and the square root of three, two of the Pythagorean incommensurables. These atoms combine to make the four regular solids with triangular or square faces, the molecules of the four elements, earth, air, fire, and water. The fifth regular solid, with pentagonal faces, was assigned to mark off areas of the universal sphere, a kind of signature of the Pythagoreans for whom the pentagon had special mathematical interest since it contained the incommensurable, the square root of five.

Perhaps the prettiest conceit in the dialogue, and not merely witty, is the introduction of the irrationals, or incommensurables, into the basic atoms; they were the pet secrets of the Pythagoreans, and Plato's own pet secret was the irrational principle that incorrigibly runs through human life and the realities it has to face. There is the suggestion here that Plato is wondering whether, if the mathematical irrational were reasonably recognized and dealt with, some of the tragic sense of life might become more sanely acceptable. Plato also knew that the sphere, the most perfect of all geometrical figures for him and for any Greek, the very figure of the universe for a Pythagorean, had that doubly mysterious irrational number, pi, at its heart. Pi tries to complete itself in an unending decimal, but even with an infinite series is unable to become the root of a rational algebraic equation; it is a transcendental irrational. There are many prophetic subtleties in the *Timaeus,* but nothing so movingly witty as this.

Perhaps the most revealing theme in the *Timaeus* is the central position and the suggested powerful use of ratios and proportions. The ecliptic, the circle of the Different with its orderly containment of all possible ratios in one perfect continued proportion is a whim with a method in it. The numbers of atomic triangles in the molecules of the elements are such that simple proportions will express the transformations of these elements one into another. The distances between the stars in their constellations, the motions of the planets along the ecliptic, the motions of bodies on the earth's measured surface, the phenomena of light, all these could be expressed in a calculus of ratios. The ratios themselves could be translated from discrete numbers to continuous magnitudes, and all geometrical figures could thus be numbered, provided the irrationals could find a rationale. There is in another dialogue an account of the study of this last problem by the young Theaetetus, a young mathematician who was killed in one of the battles of the Peloponnesian War. His studies were later carried on by Eudoxus, another member of the Academy, and their successes with the solution of the problem are memorialized in the fifth and tenth books of Euclid's *Elements*.[12]

But the Greeks also knew about analogies and figures of speech by which nonquantitative data could be brought into analogical correlation with the mathematical ratios. The hots and colds, the drys and wets of the elements also made continuous proportions. The theory of ratios and proportions cultivated from the Pythagoreans to the Academy became a calculus or logistic, and also a scheme of universal measurement, not only for music, but even for physiology. Although a method of demonstration and proof, though not Aristotelian in style, was equally developed, as the *Elements* of Euclid show, the calculus of ratios broke the ground and made Greek science a going concern. The latter part of the *Timaeus* is an unpedantic, almost humorous sample of the likely stories that such a method produces.

The more serious workmanlike application of the method is to be found in the kind of astronomy that led to the Ptolemaic system. Hipparchus, several centuries before Ptolemy took up the systematic task, saw the retrograde motions of the planets as possible optical distortions of cyclical detours from the main ecliptic paths. The sun showed no such retrograde motion, and therefore could be assumed to have no epicycle; the other planets, when they moved backward temporarily in their regular march were describing a cycle on the great cycle. If one epicycle did not take care of the irregularity, another hypothetical cycle could be assumed, an epicycle on an epicycle. Circular motion was the familiar and orthodox motion proper to celestial bodies and extra circles only add to the dignity of these admirable objects. Much known geometry of the circle and the right triangle went into the elaboration of this hypothetical invention, and there was considerable mathematical invention to keep up with it, notably the beginnings of modern trigonometry and a great deal of what we would now call operational analysis, what a Greek would call music, or the measurement of motions. The specification of the hypothesis to save the astronomical appearances used and developed the calculus of ratios with increasing success and considerable novelty. Perhaps this is the first time in Greek science that the internal mechanism of progress in thought was clearly discernible: the solution of one problem gave rise to a new problem, sometimes to many new problems. Slight wanderings from the proposed hypothetical cyclical path demanded a new epicycle, and new

[12] The tenth book of Euclid's *Elements* embodies Theaetetus's classification of irrationals effected by introducing the squares of the irrationals into ratios. The fifth book is built on Eudoxus's famous definition of "same ratio," which permits a rigorous treatment of ratios of incommensurable magnitudes, and is indeed the high point of Greek mathematics with respect to both abstraction and generality.

aberrations from that epicycle demanded still more. It almost seemed that a cycle gave birth to cycles—and sometimes to evanescent cycles called nodes. The moon proved particularly prolific of nodes. Perhaps here also is the first appearance of a scientific system which not only saved appearances but guided investigation, and thus tempted the scientific mind to its characteristic sin of satanic omniscience in systems.

Another parallel set of currents in this stream of thought is to be found in three men. Apollonius noted the figures that result from cutting a cone at the possible differing angles to the axis of the cone: the circle, the ellipse, the parabola, the hyperbola, and the point or straight line or pair of straight lines that goes with the cut through the apex. With the calculus of ratios applied to certain standard lines and points of reference he was able to formulate general properties of these figures. Again the geometry and arithmetic of the curves were extended, and the calculus gained in power and subtlety of application.

Diophantus saw the calculus broaden itself into formulae that included operations of addition and subtraction expressing relations similar to those of the ratios, saw also the high calculative power of these formulae, and made a laborious study of the extension of the rule of three that these suggested.[13] In other words, he caught a glimpse of the modern equation and the analytic methods for finding the unknown values in equations.

The third giant in this agitation of the waters was Archimedes, whose mind was well stocked with the mathematical lore of number and figure, and who saw those mathematicals governing movement in commonplace tools and natural objects, and conceived the possibility of the mechanical solution of mathematical problems. The results were many and fateful: the statics and dynamics of floating bodies, the heuristic idea of specific gravity, the lever, the center of gravity, and the focusing of light by reflection in mirrors. Archimedes brought the method of exhaustion, developed by Eudoxus in the Platonic Academy as a branch of the calculus of ratios and formulated demonstratively by Euclid in his *Elements,* to the point of effective application as an integral calculus to the areas and volumes of geometric figures not bounded by straight lines or planes.

To a modern, viewing and reviewing the bright brave ships that float on this great river of thought, it is almost unthinkable that they

[13] The *rule of three* as Buchanan is here using the term (and as it was used during the Renaissance) means the procedure for finding one term in a numerical proportion when the other three are known. Thus if $A : B :: C : D$, and A is unknown, the rule of three tells us to multiply B by C and divide the result by D. For an example, see Galileo, *Dialogue Concerning the Two Chief World Systems,* tr. Stillman Drake (Berkeley: University of California Press, 1962), p. 225.

should have gone aground and sat idly waiting a thousand years for the vital water to float them again. It is unthinkable that this versatile, clear, and eminently applicable mathematics should not have itself raised up the inventors and the engineers, who could have matched the architects of Roman law, with those exploits of science in technology that have made our modern world. But the science of mathematics never moved from Athens to Rome, the new capital of the ancient world. It started east with Alexander and lodged in the great library at Alexandria, where only part of it, some shining examples to be sure, survived the great fire. The fact is that we do not know the major portion of it, at least by volume of writing, and to this day the classical dictionaries do not give scholars the aid they need in interpreting the surviving texts; the large Oxford *Greek-English Lexicon* records the thorough study of the literary classics but has not yet caught up with the equally classic scientific writings of the Greeks.

Sublime Arithmetic

It appears at first that this stream of thought, like some Asiatic rivers, is lost in dry sands, or perhaps dived underground to travel like myths in the folklore and magic of the common people who did the work of the world. This is partially true, particularly among the Arabs, who probably basked in the light that is still hidden from us as we stumble historically through the Dark Ages. But this is not the whole truth. Proclus, who fell heir to the headship of the Platonic Academy in the fifth century A.D., wrote a book on Platonic theology, which is sometimes called *Theological Arithmetic*. Proclus is called a Neo-Platonist in the classification of schools of philosophy. This means that he was one of those laborers in the many attempts to give systematic unity to the teachings of Plato and Aristotle, a project that would have raised the opposition of each dialectical mind. The usual result of these attempts at synthesis was a partial mathematization of Aristotle.

Proclus wrote commentaries on the *Timaeus,* the *Cratylus,* the *Republic,* and the *Parmenides,* and then a work on Aristotle's *Physics.* The reader of great books can guess what would happen. Mathematics would take the place of logic as the frame of reference in the depths of metaphysics; physics would tend to become mathematical physics, and the Platonic dialogue that would make this possible is the *Parmenides,* in which the dialectic of the one and the many first reached its full comprehensive power. The other dialogues would supply the machinery for this sublimation of arithmetic, later canonized in the doctrine of the Trinity. The secret of the sublimative process is the transformation of the number one, merely the first term in a series of counters, first into a unit, that which is counted, then into one as differentiated from an "other," then into the common property of unity found both in the one and the others, then the unity of the series, or of any parts of a whole, then through a series of ascending and increasingly comprehensive unities of many parts, to the One which is above all division and multitude, and yet is immanent in each part and on each level. This dialectic, with eight separate internal dialectical movements, is given in the *Parmenides* as an example of exhaustive dialectical exposition. In a variety of contexts it seems to carry the whole of human opinion, science, and wisdom in one simple insight that interests and convinces minds of all ages and stages of sophistication. Unity is, of course, one of the first principles of science without which no thinking can be done and with which all thinking is done, for the most part implicitly and uncritically.

This is the dialectical device by which the heights of the Aristotelian metaphysics were stormed and taken by another Neo-Platonist, Plotinus. At the highest point of that metaphysics were the mutually reflective relations and functions of intellect and being. When the intellect acts freely and completely, it thinks truly of being; when being is at its fullest perfection, it thinks. The unmoved mover is thought thinking thought. Empty and redundant as this may sound, it is what insures the intelligibility and validity of all thought. For Plotinus, with all the force and conviction of both logic and mathematics in his soaring mind, this separation of thought and being was an intolerably unresolved disunity. With a subtle and stout mind fired with an authentic mystic experience, he reached, grasped and returned from beyond being and intellect with the unity of thought, being, and nonbeing, the One. There had been many a sophist and charlatan who could have followed the verbal ascending maze to this lofty conclusion, but there had been none with the power to convince himself and others of the insight. Plotinus, steeped in the mathematical knowledge of his coheirs to the Academy, must have reached the summit by many labors of abstraction and sublimation, but he gives no account of these, except by allusion, in his great book, the *Enneads*. In fact, he did not himself write this book. Like Socrates and his reporter Plato, Plotinus and his reporter Porphyry did not trust the written word. The *Enneads* are the lyrical impressionistic fragments of lectures delivered to a society of devotees. They induce insight, if they communicate at all.

In mid-twentieth century, when the only language of communication is politics, as it was in the early Roman Empire when Porphyry lived, one can be more sympathetic with the desperate attempts of this man, filled and overwhelmed by the science and learning in his grasp, to find the medium in which light could come into the world. His solution is, to say the least, marvelous, and somewhat to our puzzlement, effective over the following centuries. He does not talk of numbers, figures, and ratios, as most of his fellow Neo-Platonists did; it is as if he assumed that these details would be taken care of by the practical and world-conquering Romans, as in a sense they were saved and transmitted by the engineering operations of the Roman army. Plotinus chose to speak in figures of speech and almost in the numbers and rhythms of poetry, if one can trust the hidden music that sounds through the rugged and ragged prose of the reporter. For the reader's mind that is at all tuned to the themes, they are mathematical in their form and their motions, although the surface is allegorical. The only other world-moving literature of the time was also allegorical, but it was reporting the science of love. The New Testament moralizes and inspires; the *Enneads* glow and awaken the mind until it explodes into hypotheses.

The allegory is a figure of speech which belongs to a family of figures comprising the metaphor, the simile, the analogy. The family relation between them can be seen in a common principle; they all compare things by coincidence of relation; or they contain a common relation with diversity of terms. The analogical form of the comparison is: A is related to B as C is related to D. Similes and metaphors abbreviate the form; allegories expand it. It will be noted that the formula for the analogy is the same as the formula for the proportion. With certain exceptions, due to the peculiarities of mathematical symbolism, the operations proper to the analogy operate also in the proportion. Most of the techniques of modern analytical mathematics, particularly in the study of functions by means of series and tables of numbers and the techniques in projective geometry, are analogical and allegorical. Words and images have a wider range of ambiguity than numbers and figures, but this is a matter of degree, since the essential property of mathematical symbols is controlled ambiguity. It is therefore possible to substitute words and images for numbers and figures, to add certain controls that a poetic diction provides, and to talk about many things in a mathematical manner. The atmosphere is a bit fantastic when this is attempted, but so is it in mathematics until familiarity begins to breed a contempt for the commonplace. There is no doubt that both Plato and Plotinus counted on friendly discipline and even institutional understandings to support their desperate extrapolations of mathematical thought, and that the fantasy is intentional. They reestablish the perpetual ceremony and ritual of celebration. It is said that the Pythagoreans sacrificed a hecatomb to celebrate the discovery of the theorem that goes by their name; the thousand words sacrificed by Plotinus to the One are probably a better investment, since many words gained new life and passed it on to a dying world.

But the semantic novelties of the style and diction were in the service of a vision which had even more vitality and power. The Good, which Plato in one sentence in the *Republic* said was the source of all existence and value, and the unmoved mover, which Aristotle said in the *Metaphysics* was the cause above all causes and moved the world by love; both of these are for Plotinus blasphemous distinctions within the unbroken and unbreakable unity of the One. The Good and the Unmoved Mover have on them the marks of the dichotomizing intellect, and the limited powers of the discursive demonstrating reason. The One is above all such partialities and forms, and is more than all things and sums of things. It is power and perfection plus a source of abundant overflow of being, more than any possible world or infinity of possible worlds could receive or contain. It is the inexhaustible fountain from which all change and all eternity have their life. Where Plato had said that madness, poetry, and love are necessary to apprehend the ultimate things of reason, since reason reaches beyond itself, Plotinus reports a state of ecstasy, the mind standing outside itself, as the condition for even the dim recognition of the One.

For one who has tried the paradoxes and antinomies of metaphysical speculation there is great wisdom in this recognition of the limits of reason. It had often been recognized by keen minds before Plotinus, as it has always been recognized in the great religions, with or without the help of theological formulation, but it adds the clarity of recognized ignorance to have it expressed in such uncompromising and exhaustive negations. There is no false humility, such as is often attributed to Socrates, no arrogant agnosticism such as seized upon the nineteenth-century defenders of science, none of the opaque dogmatism so often charged against all of scholasticism.

Nor is this balanced admission of the possibility of ecstasy an excuse for intellectual laziness and the many dangerous vices that accompany it. The One, as Plotinus reports it, is both a warning of the excessive expectations and burdens that can be laid on thought, and also at the same time a certification that its proper uses are not futile. The uses of the intellect are to be measured, and to be transcended, if possible. The intellect receives its nature, its objects, and its discipline from the One. Truth is the business of the intellect, and it is possible, difficult, and good to achieve it.

But perhaps more important for science than this charter of speculation at its limits, the One is king in all his realms; each part of reality is one made out of many. There are those who see in this absolute and its dependents signs of the baleful influence of the Oriental despotism as it colored and qualified the rule of the Roman Empire, but Plotinus, as if he were aware of this, used his novel diction to remove his thought from any such political influence. As there is no hierarchical domination between the levels of an allegory, or between the two sides of an analogy or a proportion, so there is no external necessity coercing the parts of the world, or the parts of thought. Each nature has its own power to maintain its own being and to fulfill itself with only the limitation that comes from the narrowness of space and the slowness of time, which are evidences of nonbeing. The power of the One is transmitted and responded to as persuasion, as Plato would have called it, or as love, as Aristotle would have called it. Any effect the One might have on nature could only be an increment of being, and there are many signs of such increments in the world, enough to take care of latter-day evolution and human development. The term that Plotinus used for this effect of the One on things was emanation, a radiation like light and heat from the sun.

The apparently fixed natures of things have been the termini of the emanations, and as existing are the recipients of the emanations. These natures, which Plato would have called ideas or essences, and which Aristotle would have called forms, are again called by Plotinus a new name, a Greek word often translated *hypostases,* but better translated *precipitates.* The hypostases, or precipitates, exist in an infinite variety of kinds and in infinite degrees of reality. Greater being means more qualities, and less being fewer qualities. Out of this conception have come some of the most pervasive principles in modern science, the notions of infinity and continuity, which are to be found most notably in mathematics and its applications, and in biology as theories of evolution. It is perhaps well to recall that the conception of the One, its emanations, and its precipitates came out of Greek mathematics. It is not surprising that Plotinus should re-emanate them.

This is probably the first philosophical system, and perhaps the only one that is properly called a system. The Renaissance Period saw many copies of this one, notably Spinoza's, but there was none adequate to the essential requirements. These requirements are the necessities of mathematical thought. There must be a single central or highest unity representing the whole, an array of elements representing the parts, and a scheme of relations connecting the parts and parts of parts. The system can be seen as a hierarchy, or as a circle with sectors expanding from the center, or as a frame of reference organizing a manifold around an origin. Systems are more easily organized in special branches of science. Plato may have seen his work as the exploration of an ideal system, but his dialogues show only expeditions and episodes to key points of such a unity. Aristotle sharply delimits and defines his field of inquiry, and leaves the parts, the sciences, in unrelated separation. By such division it is possible for the kingly unity to rule, that is, to provide some interpretation for each item that relevantly belongs in the field. If the field so explored is clearly a part of something more inclusive, troublesome items can be excluded, and the system maintains its integrity. Plotinus set and accepted for his task the systematization of everything, including nonbeing. This involved the greatest possible effort in dialectic, a maximum of imaginative formulation, and in the end a full recognition of an abundant overflow of beings beyond the comprehensive power of concepts. The secret of the comprehensive power that a system maintains is an idea that has since the eighteenth century taken its central position in mathematics, the idea of function, the idea of the activity proper to a thing. It is an idea that also plays a crucial role in physiology, the functions of the organs and the function of the organism as a whole.

Any concept can be translated from its flat unity into relational and operational patterns. The items that as a unity it comprehends appear then to be stages of a continuous transformation of some original element. Thus a circle can be understood as the curve resulting from the rotation of a line-segment about one of its end-points, an institution as the result of human beings living under a rule or set of rules, a soul as the activity of a body. The last illustration, taken from Aristotle, is the rhetorical basis for the development of Plotinus's system. The One first emanates into an intelligence and then emanates further into souls, a world soul, and then into human and animal souls, as well as into star and planet souls. All souls are active functions and there are souls that are functions of souls.

The strict mathematician is shocked by this rhetoric because he knows that a mathematical function has no such unity. An abstraction has been reified, an hypothesis has been mythologized. It is the property of mathematics to take care of the elements, but not to manufacture new entities. The mathematician does not want to become a biologist or a theologian. He would prefer to be a mystic. And that is what Plotinus the philosopher is. He prefers to follow the argument where it leads. If numbers lead back to unity, and the universe is accepted (Gad, you'd better!), then there is a legitimate though dark question —What is the One? Plotinus, with many warnings that the answer is irrational, proceeds to try the answer, and it overflows with significance, enough to fill and more than fill the world, a result not irrelevant to the high aim of all science. So there is apparently a high intellectual responsibility in recognizing that the world is full of souls.

The speculative madness goes one step further. Plato never knew what to do with the ideas that are not directly exemplified or participated in by things. These came to be called the separated ideas. Many of them, such as equality and the perfect circle, occurred in mathematics; in modern mathematics there are many more, such for instance as the transfinite numbers. There are many other such ideas in other kinds of valid discourse. As pagan philosophers critically explored their own theology, and as Christian and Jewish philosophers developed theirs, these incorrigible abstractions took on reality, like characters in the drama and the novel. The speculations on them converged on a common solution, the lines of the speculation passing through the Olympian hierarchy, the Christian and Jewish parallels through the angels. In the end a simple proposition was arrived at: the separated ideas are intelligences. This became common doctrine in the scholastic period, and it will turn up in the science of the Renaissance. In this sublimated frame of reference there seems to be a deep connection between functions—mathematical, biological, and human.

This has been a long, wordy, and bewildering detour back to the Pythagoreans and through a labyrinth of numbers, figures, and ratios. But all this, and more too when the history of science does a better job of recollection, is necessary for a modern reading of Dante, who by some almost telepathic vision remembered all these things, and remembered too the ladder in the philosophy of love. Undoubtedly, one of the threads of his memory was a theme originated again by the Pythagoreans, and raised to central doctrine by the Stoics. They also had a system of sorts, a philosophy of life, as we have come to call it. In this there are only three sciences necessary for good human life: physics, logic, and ethics. The subject matter of all three sciences had a common part, the Logos, the Word, as it is called in the Gospel according to St. John. They used their science of logic to discern and to formulate the laws of nature. These laws in theoretical terms became physics, and in practical terms were the rules of conduct. It was the part of a good man to return to nature, admit and accept natural necessity, and hold for his practical end only those goods that were allowed by nature. The order in nature would point out to him the right path for his life.

This obviously evidences the deep sense of defeat that latterly marked the ancients' view of their adventures in the arts of civilization, but it also exhibits the solid achievements of that adventure. A science of nature, physics, had founded itself on the logoi of forms, numbers, figures, and ratios, the formulae by which causes could be understood. These are natural laws. But another natural law was also to be found in the inner parts of a man, the laws of his means and ends. These two stemming from the logos, the order in all ratios and the order underlying all political laws, were the intelligent fates of the whole cosmos, and together constituted the great intelligence ruling the world, Providence. It is not surprising to find it personified and deified in Christian theology as the Word, the Son of God, the second person of the Trinity. It was St. Augustine who canonized mathematics in Christian theology. In him the sense of calamity and frustration, both in his own life as reported in the *Confessions,* and in the world as reported in the *City of God,* dammed up and compressed his feeling and imagination until they found in the ladder of love and the mounting allegories of the Bible the intellectual heaven of the Logos. In one treatise after another Augustine finds the itinerant form of the logos in numbers, ratios, the figures of geometry, and the figures of speech, until by his own style of Platonic reminiscence he arrives by steep abstraction at the home of thought, the Logos. Through him the Psalms become the authority for the scholastic proposition that God has disposed all things in number, weight, and measure, as well as for that other proposition that He is the giver of all good and perfect gifts, the greatest of which is love.

Plotinus and Augustine both were artists in barbarous speech. This is so partly because of the endemic desperation of the times, but it is also true that they were loading words with themes that they had never before carried. Their successes in communication are a bit incredible. Part of the success is due to the disciplined following that each had, one in a pagan teaching cult, and the other in the Church. They were both teachers, and their rhetorics fitted their forms to the immediate human audience whose collective habits of mind condensed and crystallized the doctrines. They were voices in a conversation that later developed into the dialectic of disputation and *Summa.* They were the initiators of a technique of transmission within a tradition. A thousand years later their writings themselves in actual physical possession were almost unnecessary. Their lore had been contentiously and lovingly processed, purged of crudities by time and the disputations of the universities, so that Dante in exile could rethink and recompose both language and themes.

It would be a great mistake to try to find Dante solving problems in mathematics, optics, and music or demonstrating propositions in geometry or ethics. All these things are scenery for the drama, even discarded scaffolding for the building; only the vision remains. The scheme is three cones, two mirroring each other with their pyramided circles, and another marked with a spiral. This is the geography and the astrology of a double journey, Dante's and the world's. There are graded distributions of light, heat, and weight, with proportioned speeds and accelerations of motions. But all this is allegory for passions, virtues, vices, sights, and mysteries, which tragedies and comedies reveal, in the music of number and verse making one intellectual world. Intelligence meets intelligible in echoing lights and sounds. This is probably the only time that scientific method has transcended its means and seen its ends.

The Dream at Work

Nevertheless, the typical man of the Renaissance sees illusion in this vision. It is as if Dante were a bright angel who had brought a dream to a world that had long been asleep. The Renaissance scientists reacted to Dante as if they were psychoanalysts who knew that men had originally fallen asleep because they were not able to face their frustrations, their failure to put mathematics to work in the useful arts, to spell out their insights in the affairs of the world, in short to bring off the industrial revolution. Dante's poetry is merely the vehicle of a compensating dream work to keep men from waking up. So the Renaissance man, even in the monasteries, slowly and dreamily sets to work finding the light in the lowly water wheel, the windmill, the fulling mill, the grist mill, the sailing ship, the mechanical clock, the lever, the pulley, and the screw. Some of these are ancient, some of them new or renovated. The mechanical principles are fairly familiar and recognizably exhibit the traditional formulae composed from numbers, figures, and ratios. The new men steal from the dream, and embed ideas in things. It is said that the monasteries were the first human institutions that rotated handwork, brainwork, and spiritual exercises daily for each brother. The dignity of labor, of man, the fabricator, and the beauty of the machine are the phosphorescent reflections of intellectual light. But the light itself was disclaimed even when it was taken for granted. The dream was a symptom of a deep trauma.

Some of the other symptoms of the trauma took the form of attacks on the content of the dream. The scholastics who had helped to compose the dream were called logic-choppers, refined sophists, and exhibitionary elaborators of the obvious. These attacks were explicitly directed against Aristotle, who not many years before had been the pagan infidel, then merely the philosopher, and finally The Philosopher. Aristotelians were personified as stupid fumblers and doctrinaire snobs with respect to natural things. Theology was branded as vain speculation. Dialectic was continued in the universities, but it was discredited as hairsplitting verbiage. To a modern this is confused with his own anti-intellectualism and what has been called the treason of the clerks, first portrayed in Erasmus's *In Praise of Folly* and in Rabelais's *Gargantua and Pantagruel,* and now expressed in sophisticated revolts against the pedantic scholar. All this can be attributed to the spontaneous reaction against authority that has outlived its validity and to an enthusiasm for doing things naturally, but it does not throw much light on the revolution in science that follows.

No full searching light can at present be thrown on this period of intellectual history, not even by those who ignorantly suppose that a pure genius like Leonardo was born out of the head of a pagan god. The Renaissance man was filled to overflowing and bursting with the content of the dream. He was pious, intellectual, and even traditional. With all his bluster and bravado, he was shrewdly honest. He thought he saw a trick, and he wanted to correct it. Just what that trick was and what he thought he saw to correct in it is one of the mysteries.

The unknown in this mystery is connected with our historical hypotheses about the Dark Ages. It is more than likely that we call them Dark because we do not penetrate them. In 1950 it is salutary to remember that there have been Iron Curtains before our time, opaque and fearsome barriers between cultures. Iron Curtains leak; they let through the finely sifted dust that glows and blinds us. It is more than likely, from what little we now know, that the Dark Ages are a shadow cast on our own past by the iron curtain we erected against the Moslem. It is fairly clear that the knight-at-arms and the code of chivalry sifted through; it is even more clear that algebra filtered through and saturated our minds. The Arabs had also been vehicles of the ancient lore of love and mathematics, but they had added something of their own, a ritual of love, and a symbolism for all of mathematics. Out of this we have the comic account of chivalry in Cervantes's *Don Quixote,* Rabelais's *Gargantua and Pantagruel,* Erasmus's *In Praise of Folly,* and the eighteenth-century father of novels, Fielding's *Tom Jones.* All these are not only comic; they have the steady determined bite of satire, and they are obviously parodies patterned on some tragic and sublime theme. They are saying that something important has gone wrong and only laughter will set it right. In Descartes's *Discourse on Method* there is an undisguised frontal attack on the lumber and machinery of the scholastic disciplines, a purgation by doubt, ending in the cool formal traditional proof of the existence of God. The rest of the discourse outlines a method in which clear and distinct ideas, intuition, and deduction are substituted for the discarded apparatus of scholasticism. This new method is taken from the techniques of algebraic analysis.

The Analytic Art

We do not know enough about Arab theology and mathematics to attribute the revolution to the Moslems, but we can go back to the common source and make a shrewd guess concerning the origin of algebra. The Greeks had distinguished between two kinds of demonstrations. In one, called synthetic, known premises were formulated and combined to issue in a conclusion; this is the familiar procedure with common notions, definitions, postulates, and hypotheses, with a Q.E.D. at the end. But it was noted in this that the Q.E.D., the thing to be proved, acted as a guide to the proof, not only ordering the steps of the proof, but requiring now and then constructions—parts of the figure which were not given, but invented to fill in missing steps in the proof. It was then brightly presumed that a really ingenious geometer might begin with the conclusion, invent all the steps of the proof, and work back to the given premises. This involved inserting certain hitherto unknown terms and uncertified steps in the proof. The theorem to be proved acted as a regulative form for such invention and improvising: the form does not create or give birth to the invention, but acts as a rule or criterion by which the trial construction can be judged relevant or in error. When these constructions connect, as some of them will, with the known premises or the given facts, the theorem is proved. This kind of proof, which incidentally comes closer than ordinary demonstration to the original meaning of proof as a test of validity, is called analysis. It can be used in proofs of arithmetic, as well as geometric theorems, as is seen in Euclid's *Elements,* Books VII, VIII, and IX. It is used most swiftly and effectively in *reductio ad absurdum* arguments in general.

Algebra came into Europe as the analytic method, probably at the time when many Jews and Arabs were imported to teach the hordes of semibarbarous students who flocked to the new universities. Algebra was bootlegged knowledge, giving many new tricks for solving difficult problems in arithmetic and geometry. It can be acquired for the easy solution of simple problems, as it is often taught to high school students today. Given certain known terms, usually arabic numerals, or sometimes the first letters of the alphabet, one is to find the values of certain unknown terms, expressed by x, y, and z. The solution of such problems proceeds by carrying out the operations indicated by the symbols of addition, multiplication, subtraction, and division which connect the symbols representing numbers and magnitudes in an equation. The equation is the theorem that guides the "construction" of trial values, and the operations seem to deduce the correct answers. The mathematician, who uses these formulae without tears, knows that the equation had to be discovered or invented, and that the rules are merely forms of mathematical inference based upon the four fundamental operations of arithmetic.

It is often said that the dullness of algebra for the ordinary student who is forced to take it in school comes from the pedantry of Arab learning in general, that the bright side of the Dark Ages was just a dull glow of imitative rules and operations. But the evidence we have from the effect of this learning on the early European mathematicians and philosophers does not bear this out. Descartes, and several other competent and brilliant minds before him, saw visions in the symbols, and these visions led them to further invention, sometimes inventions of symbols to facilitate the new art—decimal fractions by Stevin, coefficients by Vieta, and powers by Descartes himself. But Descartes also saw deeper. He saw the equation bridging the gap between numbers and figures, or discrete and continuous magnitudes, as the Greeks had distinguished them, and this led to the invention of analytic geometry.

Pappus reported that the relative distances of a point from two, three, or four lines could be stated in terms of proportions, and that when these were so stated the formula gave coordinates for a locus of points on lines or curves, but that the problem of the coordinates to five lines had stumped him and his fellow mathematicians. Descartes was able to state the problem in terms of equations, and to show that these equations governed the loci of all the conic sections. These are quadratic equations which transform the relevant calculus of ratios into the family of equations of the second degree, those that contain squares. Equations of higher degree express loci for other curves which have family characteristics. By referring all these to two lines intersecting at given angles for two variables, and to three lines at given angles for three variables, he saw the possibility of generating equations to fit any locus, and thus of reducing the world of all geometric figures to one system of algebra.

As one reads the account of this discovery in Descartes's *Geometry,* it is very difficult to appraise its intrinsic importance. It is news in the history of mathematics, but as a critic said at the time, all of it might have been learned in the work of Vieta, who had done most to give mathematics its most characteristic form of expression, the equation. For one who has read the Greek mathematical classics, it is almost impossible to believe that the Greeks could have failed to solve the problem of Pappus, since there seems to be no part of Descartes's solution missing in their calculus of ratios. From the modern point of view, a proportion is an equation. Perhaps right here is the implicit secret: for the Greeks the ratio was a radically different mathematical category from the characteristic operations and symbols of arithmetic and geometry; in Euclid the fifth book opens up a new treatise, quite independent and quite removed in degree of abstraction from the rest of the book. Furthermore, the ratios of arithmetic dealing with discrete quantities are quite different from the ratios of geometry which deal with continuous quantity. There was no rationale for fractions in Greek mathematics, so that our association of fractions and ratios was not dreamed of.

It therefore seems that Descartes's ignorance, or perhaps willful ignoring, of these Greek distinctions, allowed him, or forced him, to push his head through a traditional shell, and to discover a new level of abstraction, to which the work of the Greeks and the work of Descartes's more immediate predecessors, Oresme, Vieta, and others, quickly and completely assimilated themselves. Like heavy charges of electricity which had been building themselves up in the separate fields of mathematics, they rushed together and discharged themselves through a single conductor, the symbolic form of the equation and its counterpart, the locus of points and correlated numbers. In this discharge there was a flash of comprehensive light, in which some of the clarity and distinction of arithmetic, geometry, and the calculus of ratios are lost, in which also some of the so-called mechanical problems incapable of precise mathematical formulation appear to be soluble in principle, and in which the intellectual grasp on function is improved. It is difficult to appraise the vision, but there is no doubt that the consequences for mathematics and the other sciences are still great and still imponderable.

The immediate effect on Descartes's mind is wonderful too. In a combination of modesty and pride, which is characteristic of the period, he recognizes and admires his own personal achievement. At the same time he is overwhelmed by the prospect of its future consequences, and as if to conceal, even from himself, its full meaning, he proposes that his discovery and invention is the merest glimpse of an ancient secret art of which the Greek records are only a vulgar report. Descartes pictures himself as the modern Prometheus stealing divine fire and exposing it for every man to possess and use. Each human mind has an aspect of divinity in it which can be disentangled, revealed, and used, if only it will submit to the analytic regulation and discipline. The ancient secret art is of course the art of analysis, and algebra is its modern prophet.

But the art is not merely the universal science of mathematics; in fact, the importance of mathematics in Descartes's view is chiefly as a discipline to train minds in a new method by which wisdom, the unity of all the sciences, can be attained. The infinitude of figures, as systematized in analytic geometry, is not only adequate to all sensible differences, but it is adequate as a set of tools to construct the images of all objects whatsoever. Part of this claim is based on a book ascribed to Nicolas Oresme, *The Latitude of Forms,* in which all variations in qualities are reduced to geometrical diagrams, a very early anticipation of the systematic reductions that have since been carried out by mathematical physics.[14]

But there is another part of the claim that Descartes himself began to carry out. In this part qualities and other nonquantitative things are not reduced to order and measure, but are dealt with according to a more generalized method in which simple verbal terms are isolated, understood, and used as starting points for a chain of reasoning which he calls deduction. Many things, he says, besides numbers, figures, and ratios, can be grasped by intuition with certainty and have clearly traceable consequences. In the *Discourse on Method* and the *Meditations* there are some examples of this method. By a semidialectical series of doubts he arrives at the clear recognition that he actually does doubt; this means that he thinks, and therefore that he exists. Or, his own imperfection gives him the idea of a perfect being; a perfect being cannot lack existence; therefore God, who is that perfect being, exists. Since God exists and is perfect, he must be the first cause, or the cause of all causes; but his perfection involves perfect goodness; this goodness precludes any deception in His relations to his creatures; therefore God certifies the validity of clear and distinct ideas. This is a brief account of the actual steps of the Cartesian argument, but in relation to their originals in traditional scholasticism, which he is condemning, they are a little like childish diagrams, or caricatures. But they are fairly good examples of the method, clear in conception, hard to refute once the original insight is apprehended, and powerful in a systematic context. They are typical of much modern thought that has followed the controversies that the Cartesian method initiated.

In the *Discourse on Method* Descartes gives an autobiographical account of his reaction against the scholasticism in his early Jesuit education, his discovery of the seeds of a new discipline in the analytical art

[14] The little book *De latitudinibus formarum,* with the author given as Oresme, was printed at Padua in 1482 and 1486, at Venice in 1505, and at Vienna in 1515; the wide distribution of copies in European libraries suggests that it was the main source through which Oresme's doctrine was known in the sixteenth century. However, its probable author was not Oresme but Jacob of San Martino (see Marshall Clagett, *The Science of Mechanics in the Middle Ages* [Madison: University of Wisconsin Press, 1959], p. 395). The much lengthier and more innovative *De configurationibus qualitatum et motuum* by Cresme was never published during the Renaissance, but has recently been edited and translated by Marshall Clagett in *Nicole Oresme and the Medieval Geometry of Qualities and Motions* (Madison: University of Wisconsin Press, 1968).

of algebra, and the project of intellectual reconstruction that his counter-conversion demanded. Some see in the *Discourse* an imitation of Augustine's *Confessions,* and the comparison that this suggests is interesting. In place of Augustine's recollection of sin there is the Cartesian purgation by doubt; in place of the allegorical method of reading the Scriptures, which Augustine learned from the sermons of Jerome, there is the analytic method which Descartes learned from Vieta and the Arabs; in place of faith in divine truth there is the intuitive grasp of clear and distinct ideas; in place of understanding and charity, there is the rational ordering of knowledge. There is caricature in this imitation, but there is little humor. If the original that is being caricatured is to be found in Dante where Christian faith and understanding are fully explicit, we have to add Cervantes's *Don Quixote* to complete the caricature. The theme in Don Quixote is love, but the method in the madness is analytic. The Don is the embodiment of intuition and deduction. Caught by a difficulty in the code of chivalry he proceeds by an intuitive leap to the formal solution of his problem. Dulcinea is a fair lady for whom the Don must vindicate his devotion. He therefore goes forth on adventures. In the system the windmills are giants, and Sancho is the governor of an island. Any lover or reformer knows that this is the way the world becomes intelligible and gets set right. The adventures become a system rationalizing the passions of the soul and taking the place of the ancient science of love. The consequences of this system are also imponderable, since they are still developing in the novel and in psychoanalysis.

The puzzle of the Renaissance as the rebirth of ancient learning is partly solved by a recognition of the Christian matrix; the rest of the puzzle is given a shadowy solution by recognizing the evidences of Moslem paternity in the art of algebraic analysis and the code of romantic love. Both of these are heretical in the radical sense of that word; they tear out shining threads from the luminous orthodoxy of the medieval intellectual world. In Descartes's system, which is the result of his reconstruction according to analytic method, one of these heretical parts, algebra, reflects an underlying substance which he calls extension. In this substance, equations specify the corporeal forms in nature. The other heretical part is called thought or mind, and within it there are to be found reason, passion, and sense, which are analyzed in their own terms with almost mathematical precision. Both extension and thought are independently subsisting substances, with a third substance, God, existing as original cause, but acting on the two lesser substances only on occasion to effect miracles. There is a deep meaning in Laplace's remark, two hundred years later, that as the mathematical analyst for a complete system of the world, he had no need of the hypothesis of God.

This is the revolutionary consequence of Descartes's claim that the infinitude of figures is adequate to the diversity in things. Mathematical analysis, taking the place of theology, has become the queen of the sciences. A parallel statement in the science of love says that the brotherhood of man does not imply or presuppose the fatherhood of God. The One of Plotinus is dissolved in the analytic system that orders the many. There are points in Moslem theology that harmonize with this, although the doctrine of method and system would be heretical with respect to that orthodoxy also.

In terms of the three parts of science, and the methods that result from successively subordinating two parts to a third, the methodological picture of the Renaissance is simplified, in fact oversimplified unless the ancient complication and the medieval crisis in speculation is accepted as background. As the ancient world had seen and formulated principles, exploiting fact and hypothesis for this end, so the Renaissance scientist took principles for granted and sorted and manipulated facts for the hypotheses that they would project for him. It is not too much to say that the intuitive induction that Aristotle had described was for the Renaissance a predominant habit of mind. For this habit there were several supporting conditions which the Greeks did not have. There were new tools and machines of the useful arts that embodied many mathematical and mechanical forms. These were turned more and more to the uses of speculation. Such scientific instruments keep their significance for speculative purposes only if there is some symbolic form in which their findings fit. The new analytic art of algebra, with its omnivorous appetite for figures and numbers, supplied that symbolic receptacle. It also extended and facilitated the calculative and measuring techniques. Intuitive induction under such conditions plays a catalytic role. It contributes to the invention of new instruments, and also forces the invention of new symbols for the analytic techniques; mathematics and machines consequently multiply and proliferate. There were many Leonardos who lived busy lives in seeing, calculating, and manipulating, and the things they saw were for the most part similar to the half moon of Aristarchus.

The Century of Genius

A. N. Whitehead in his *Science and the Modern World* has called the seventeenth century the century of genius. The men whom he calls geniuses, Galileo, Gilbert, Harvey, Kepler, Descartes, and others, along with the earlier Copernicus whose revolution of 1543 by some kind of delayed ignition sparked their minds, have often been called the founding fathers of modern science, as their work is now recognized as the theoretical foundation of modern technology. The foregoing account of the birth of this brave new world does not dim the luster of the heroes; in an incomplete and fragmentary way it introduces the tradition and the environment as some of the conditions for the brilliance of their performances. If the record were more complete and more continuous, the century with its heritage and its heroes might better be called the century of the great hypotheses.

In a very remarkable way the work of the ancients reached a kind of intellectual limbo in which figures, numbers, and ratios, shining with an inner light, wandered in a gloom which prevented full mutual reflection, but enough diffusion of light to suggest the dawn. The dawn was announced by Descartes in the method of universal mathematics that algebra seemed to be. This made an intellectual community. Some of the heroes show no signs of recognizing or accepting the revelation by Descartes, others observe and infer as if they have had a dim vision of the end of the new method, but prefer to work toward it with the old style and method, and still others see both the ends and the means and successfully realize the full vision.

Descartes himself, in addition to the exposition and development of analytic geometry, dabbled in physics and physiology. His major physical hypothesis, the view of nature as the locus of all possible geometrical figures, is little more than the matrix for the operation of the analytical art. The essential starting point for all the later developments in analytical mathematics is there, the assumption of continuity and infinity which later both revised the number system and forced the search for non-Euclidean geometries, but this is promise to be fulfilled, rather than accomplishment. His attempt to articulate the field of a new physics by the application of a world equation which would subsume all other equations and show that all motions are vortices or parts of vortices was a legitimate try, but it issued in a suggestive allegory rather than in the universal metric, which is still being sought by more laborious routes. There is still the demand for a unified field theory, not too distant allegorically from Descartes's vortices, but the efforts to achieve this are still spiralling around a notion in Einstein's mind. Descartes's simpler analogical hypothesis that animals are machines is still a regulative idea in physiology, but it has not yet found the adequate mechanical model.

This is not to belittle the Cartesian achievement, but it is to point out that some hypotheses live and move in various ways for long periods of time with very little clarity of definition or support in fact, provided they discharge the functions of leading principles with respect to other hypotheses. Thus the great Pythagorean hypothesis that everything is made of numbers, transformed into the Cartesian hypothesis that all events are values in algebraic equations, inspired and disciplined the century of genius. It has also continued to raise the questions and extend the age-old discussion of the place of mathematics in science. Thus the Cartesian hypothesis claims and exerts the power of a principle.

That the hypothesis itself is vague and ambiguous for all its appearance of clarity and precision is evidenced in the considerable variety of meanings that it has in the hypotheses that appeared and worked in the field that it defined. Galileo and Copernicus are dramatically opposed with regard to the use they made of the mathematical formula. They are mathematical investigators who hardly seem aware of the new analysis, not merely because they precede Descartes and his definitive formulation of the new method, but because they are artists in the old techniques with ratios and proportions.

On the other hand, their thought moves in a medium and spirit of the new mode of significance. Copernicus goes back to the original alternative hypotheses that Ptolemy considered in the opening pages of the *Almagest.* Ptolemy was quite aware that the geocentric assumption for his epicyclical hypothesis was only one of many assumptions. After a thousand and more years with all the complications that its development accumulated, it appeared to Copernicus that the alternative heliocentric assumption might be fruitful if it were applied to the Ptolemaic data.[15] It is often said that improved observations of the planets were refuting the Ptolemaic hypothesis, that there were crucial data that the Ptolemaic hypothesis could not explain, but that can hardly be the case, since epicycles and eccentrics together can in principle be supplied for any irregularity. It is also often said that the simpler the hypothesis the nearer the truth it is, and that therefore the increase of complication in the Ptolemaic system decreased its probability. Although it is true that precision and rational certainty in the Ptolemaic system left something to be desired, as Copernicus says, he seems to be putting more emphasis in his own mind on the improvement of mathematical style and method in his own proposed hypothesis than on the empirical and pragmatic failings of Ptolemy. His famous revolution is in principle no

[15] Contrary to an oft-told tale, here apparently accepted by Buchanan, Ptolemaic astronomy did not undergo much increasing complication in the centuries between Ptolemy and Copernicus. Predictions of planetary positions in the first half of the sixteenth century were still being made from the Alphonsine Tables, which were purely Ptolemaic in almost every detail. However, a supposed irregularity in the precession of the equinoxes discovered by Arab astronomers did influence Copernicus, providing one of his reasons for attributing motion to the earth; see J. E. Ravetz, "The Origins of the Copernican Revolution," *Scientific American,* CCXV (October, 1966) pp. 88–98.

more than the invention of a new perspective in painting, a new scheme for seeing, understanding, and ordering the familiar objects of a natural scene. It has enormous consequences, as no doubt there were subtle and far-reaching consequences of the new perspectives that painters were discovering and inventing at the same time. As usual, there were second-rate minds who did not distinguish between hypothesis and dogma, and many of the consequences of the heliocentric theory occurred in perturbations of their prejudices.

At any rate, with the sun as the center, much of Ptolemy, including circles and eccentrics, as well as the store of Ptolemaic observations, was easily translated to the elegance of the Copernican system. The ease and fluidity of the new analysis was not used directly for the most part, but the new significance that this art gave to the figures, numbers, and ratios of the old art had the effect of establishing a new science of astronomy, one which in the sequel has proved to have great versatility and power of assimilation and expansion. These somewhat secondary advantages originate from the intrinsic excellences of the new science as a system.

But Copernicus makes much more of a related advantage. He often speaks of the certainty of his system as against Ptolemy's. This at first has the familiar ring of the scholastic imitation of deduction from self-evident principles by Aristotelian formal logic. On second reading Copernican certainty has quite a different meaning, which again comes from the systematic virtue. He seems to mean that the application of his hypothesis to the data renders unambiguous and clear judgments of the facts. The unification of the manifold data does not strain and force the facts; it adds to their clarity and logical force. There is no appeal either to principle or to fact to bolster or certify the conclusion. The conclusion flows freely from the hypothesis, clearing rather than blurring the vision of the scientist. This is like the arguments concerning the new perspectives in painting of the time. The new projections of three dimensions on the two-dimensional canvases made it unnecessary to place ornamental vases in positions where other perspectives would have produced disturbing distortions, or in photographic terms, images out of focus. The Copernican hypothesis, like many others of the time, acts like an improved lens with large scope and finer definition. This is what Copernicus means by certainty.

The contrast between Copernicus and Galileo within this intellectual community formed by the new method is perhaps best symbolized at this point. Where Copernicus is interested in the clarification and strengthening of intellectual vision, Galileo made a telescope to extend and clarify ocular vision. With full use of the laws of ancient optics and an empirical grasp of the refracting powers of glass, he designed and ground the lenses that brought the moons of Jupiter into an ocular field for the first time, and that allowed the same knowledge of optics to interpret the clearly focused vision of the face and limbs of the moon. The thorough study of details together with the concentrated application of several intuitive inductions upon the subsolar system around Jupiter confirmed Copernicus's views by analogy, and introduced the instrument and the mathematical reasonings by which the Copernican system was to conquer progressively the field of astronomy. The report of Galileo's work is given in miniature in the *Sidereal Message,* the model for all scientific monographs written since his time.

One of the controlling allegorical figures of the time, so permeative that one can assume its implicit presence even where it is not explicitly alluded to, is the mirrorwise correspondence of the microcosm with the macrocosm; the great world is mirrored in the small parts of it. Although this image has an ancient origin, the telescope demonstrates it clearly and almost literally. The human mind is usually assumed to be the perfectly reflecting small mirror, but almost any scientific instrument, starting as an aid to such reflection, tends to develop into a substitute for the mind. Galileo's refracting telescope, whose construction is described in the *Sidereal Message,* was an auxiliary eye. It is not only a system of lenses magnifying the field of vision; it had a mechanical mounting which rotated and raised and lowered the axis of vision, and an internal adjustment for clear focussing. So far the analogy is merely imitating the eye. But by calibrating the arcs of rotation and ascension, the telescope finds itself identified with the ancient astrolabe which could be used to calculate mechanically the position of the planets and stars and also to tell time. The astrolabe was already a microcosm for both terrestrial and celestial spheres, and a measure of their relative motions. To this the telescope finally added the Cartesian axes of reference and coordinates as crosswires made of spider's web, a symbol and instrument of search in a world of meridians and parallels. If we anticipate history from this point, we can see the reflecting telescope and the camera completing the substitution for the human eye and part of the human mind. A juster description is that the essence of the analytic method is built into the medium of vision, and the induction from the vision becomes almost the work of the automatic machine. The observatory as an institution takes the place of the monastery as the place of contemplation; it is a model of the intellectual community which the analytic method created.

But Galileo was taken and possessed by another aspect of this method and system. It was as if the telescope were an umbilical cord by which the celestial matrix fed and nourished the infant machine. Galileo was the midwife watching, and aiding and reporting the first motions and utterances of the offspring. To be sure, there had been machines before Galileo's time, but they had been small, feeble, and dumb organs of an organism not yet dreamed of. Galileo, like Leonardo, went about his mechanically ingenious world, starry-eyed and eagle-eyed. He picked up mechanical organs like the inclined plane, the water clock, the pulley, the cogwheel, and the pendulum, and measured motion against time, expressing the results in terms of ratios and proportions that begged to be cast into the form of equations—equations of uniform motion, of accelerated motion, of uniformly accelerated motion, of circular epicycloid motion, and the regularly irregular motions of the pendulum.

Most of the observations in this system of motions were done after the mathematical expressions for them were worked out to clarity and Copernican certainty. It may be that the chandelier in the cathedral suggested the study of the uniform acceleration of the inclined plane, but the diagrams of Nicolas Oresme must have danced around the chandelier, as Galileo counted his pulse and assumed that its rhythm was not accelerating. For Galileo, visual phenomena, falling bodies, the flight of projectiles, the swinging of pendulums, and the engineering operations in Venice were precipitates of mathematical formulae, and their workings were the visual solutions of problems in ratios and proportions.

The *Two New Sciences* was concerned with local motions and the strength of materials, shadowy anticipations of the laws of mass and force. These are the elements that demonstrate that machines are models of equations, and that the laws of mechanics and dynamics are the backbone of modern physics. As in the case of the mechanical clock, for which Galileo had completed the design but did not supply the actual model, so his *Two New Sciences* did not give the canonical formulation of the laws of mechanics, although the terms and elements were in his hands. His work was the application of a method rather than the full recognition of a system. This is almost pathetically confirmed by his reaching for the astronomical work of Kepler, which just exceeded his grasp. In spite of the sublime precision of his first visions and interpretations of the moon, Jupiter, and Saturn through the telescope, his interests lay in the mechanisms that have been incorporated in the mounting of the telescope, for most of which he made the crucial discoveries and analysis. He was more certain of the local motion of the earth than he was of the harmonies of the orbits.

Another pair of heroes bear somewhat the same relationship to each other that Copernicus and Galileo did: Gilbert and Harvey. Each of them worked on phenomena and with techniques that are considerably removed from astronomy and mechanics, where equations were being applied directly to bodies in motion. In fact, their thought and their diction seem innocent of the new analytical art. On the other hand, any attempt to see through their writings to the unity of their thought or to the exhaustive continuity of their searching operations would miss the crucial turning points and the termini if it did not start with the confused view of the phenomenal whole and watch the points of clearing and articulation that appear within it. The whole which governs its parts and in turn responds to their variation is the rough image of the system and the method of the analytic art. Like Copernicus, Gilbert without ignoring the parts tends to the systematic view; like Galileo, Harvey tends to the ticking of the machines.

Gilbert's controlling image is the microcosm and the macrocosm exemplified in the earth and the lodestone, the *terra* and the *terrela,* as he calls them. The thin connection between them at first view is that they both point toward Polaris, the lodestar. The fascination and excitement of such a large and mysterious shadow of an hypothesis did not prevent or divert him from the faithful laborious work of methodical investigation. Beginning with the then newly familiar properties of the lodestone in the navigation of seagoing ships, passing easily to their attractive and repulsive relations to each other, varying their relative sizes until finally the analogue of the earth and a compass was before him in model, he established the fundamental properties of polarity, the permeation of magnetism through iron and the focussing of the attractive power through iron armors, and a pattern of variation through a system of relative positions of one lodestone to another.[16]

The modern empiricist and experimentalist misses the quantitative expressions of Gilbert's results, and is tortured between admiring the primitive scientist fumbling so expertly with now familiar data, and disgust with the soft fuzzy words that he uses in place of hard numerical results that we now have. In place of them there is a quaintly charming poetic terminology, at first merely suggesting that these lodestones have an attraction for each other akin to love. The language becomes more literal; there is coition between lodestones and a kind of propagation of themselves in metals that show affinity. Finally, the doctrine of natural love acquires a new image in the loves of all lodestones for the earth, and there is a suggestion of a wider community of love within the celestial sphere. Gilbert hardly reaches a Plotinic doctrine of emanation and precipitation, but the feel of the system together with our hypotheses for electromagnetism prepares our understanding for Gilbert's own conclusion: the force of magnetism is the soul of the world. All that is necessary here is the precipitation of the right equa-

[16] Concerning the "iron armors," see *De magnete,* Book II, Chapters 17–22.

tions to enclose the system, but the quasi-animate force remains with complacence in the realm of the philosophy of love. It is interesting to note that Faraday, like Gilbert breaking new ground in his researches in electricity, also clings to the allegorical modes in the record which he bequeaths happily and successfully to his mathematical scribes.

By way of counterpoint, it seems, Harvey analyzes the operations of the living organism in terms of a machine. As in Gilbert, there is enough thoroughness and precision in the investigation to support a strictly mathematical treatment, and wherever the thread of his discourse needs quantitative terms, they are used. But the analogy of the human heart with a pump is the pattern of the analysis, and the circuits of the hydraulic system save the appearances without the benefit of mathematics.

Partisans of the third method in science, that in which principles and hypotheses are used for the determination, collection, and ordering of facts, are very plausible in their claim that Harvey is the hero of their school of thought. The careful observation and strategic experimentation that he clearly and triumphantly reports are good evidence for their thesis, but they are embarrassed by the theme which rises to eloquent emphasis at the end of *On the Motion of the Heart*. They accuse him of incomplete emancipation from the dogmatic influence of traditional authority when he quotes Aristotle with approval to the effect that the heart

> is the first to exist, and contains in itself blood, vitality, sensation, and motion before the brain or liver are formed, or can be clearly distinguished, or at least before they can assume function. The heart is fashioned with appropriate structures for motion, as an internal organism, before the body. Being finished first, Nature wished the rest of the body to be made, nourished, preserved, and perfected by it, as its work and home. The heart is like the head of the state, holding supreme power, ruling everywhere. So in the animal body power is entirely dependent on and derived from this source and foundation.[17]

It is true that this summary statement is demonstrated in detail in the careful observation of the development of the fetus of the chicken, reported in *On Generation*; the beating heart appears first as a function-

[17] Harvey refers to this Aristotelian passage (which is from *De partibus animalium,* Book III) in Chapter 16 of his *An Anatomical Disquisition on the Motion of the Heart and Blood in Animals;* see *The Works of William Harvey,* trans. Robert Willis (Annapolis: The St. John's College Press, 1949; a reproduction of the Sydenham Society edition of 1847), p. 74. But Harvey does not entirely agree with the passage; see his *On Animal Generation,* in *The Works of William Harvey,* pp. 378–79.

ing structure and seems to make the other organs. But the passage quoted has another significance. The comparison of the heart with the princely head of the state, or with the sun as the vital center of the solar system, alluded to elsewhere, supplies the perspective which shows Harvey as the wielder of a method and the builder of a system. He is seeing the heart and the blood vessels as a hydraulic system within which there is rhythmic and circular motion. This is a machine that emanates life by distributing blood.

Harvey was a medical man, learned in the Galenic tradition. One of the first theories of the distribution of the blood was based on the analogy of the animal body and a curious Greek implement for carrying water, the clepsydra, or secreter of water. This implement was a hollow ceramic cylinder with a sievelike bottom with many holes. It was closed at the top except for the hollow balelike handle with an opening that would fit just under the palm of the hand that carried it. The device was dipped into a cistern or spring, and as it filled through the holes in the bottom, the displaced air expired through the hole in the handle. The carrying hand closed the hole while the water was carried, and the vacuum retained the water in the vessel. On reaching the destination, the water was allowed to run out at the bottom, and air was sucked into the hole at the top.[18] This was a model of the heart and lungs. The flexible chest was the prime mover instead of water pressure, but the nose served the function of the hole in the handle, and the blood was supposed to enter the heart as the air left the lungs, and to leave the heart as the air returned to the lungs through the nose. Galen, who knew more about the mechanical functions of the heart and lungs, revised this theory of the prime mover, but elaborated rather than replaced the ebb and flow motion of the blood throughout the body, as it was implied in the original analogy of the clepsydra. The organs were swamps through which the blood flowed and reflowed in a tidal rhythm. This is the ancient theory which Harvey inherited.

The revision that Harvey made of it consisted in seeing the movement of the blood in terms of regular circular motion, as a circulation. If this reminds one of the circular motions of the celestial bodies, one will not be far wrong. In fact the Harveian circulation nicely reflects the Ptolemaic system as if in midstage of its transformation into the Copernican. The blood starts from the left ventricle on the major cycle by way of the aorta, returning to the right auricle by the *vena cava*. On

[18] For the description of the clepsydra here given, see Pauly-Wissowa, *Real Encyclopädie der classischen Altertumswissenschaft* (Stuttgart: J. B. Metzler, 1921), XXI, 807. One of the classical sources will be found in Aristotle, *Problems*, tr. W. S. Hett (Cambridge, Mass.: Harvard University Press, 1961), I, 355–59 (Problem XVI.8).

its way it has divided and subdivided its stream into many subsystems of circulation or epicyclical circulations. These subdivisions become so small and fine that for Harvey's unaided vision they vanish; he did not know about the capillaries.

It is a shame that he did not know about the capillary systems, in which these tiny vessels seem to generate and destroy themselves according to the supply of blood meeting the local organic demands, the heart itself responding to these needs and thus "making" its own channels. This modern microscopic finding confirms Harvey's gross observations on the fetal circulation of the chicken, and this in turn helps to explain the tidal movements of blood in the depths of the organs. The pulmonary circulation repeats the larger circulation in miniature, thus giving another analogue to the correspondence of microcosms and macrocosms, which is immanent in all of Harvey's thinking.

This is the matrix within which Harvey traces all the mechanical hydraulic connections, the cavities in the heart, the valves in the heart and veins, the pulsing of the arteries as the pressure wave from the heart moves along them—with the regular cyclical or periodic order of their operations. It should be recalled that Galileo checked the periodicity of the pendulum by this regular physiological routine. The mechanical elegance of the circulation is crowned with determinations of the total quantity of blood and the rate of its movement through the circulatory system. The empirical inference matches the beauty of the systematic construction.

If the intellectual history of the Renaissance and the century of genius is at all properly construed by referring it to Dante and Descartes as if they were axes of reference, the one fixing a line of vertical ascent through the forms that represent degrees of being, and the other spreading a horizontal net of proportions knitted together in one system of equations, then the resulting community of thought with its elegant articulations and its precise penetrations of nature should show increasing strains and stresses as time and work goes on. The pairs of complementary opposites, Copernicus and Galileo, Gilbert and Harvey, in spite of their harmony of aim and the echoing of their separate lights, do show this strain. Although they all respond to the new analysis, they use different symbolic mediums, mixing allegory and analysis in varying proportions and in varying degrees of penetration. Sometimes they run to the extremes of speculative machines and speculative loves with the comic results that the more literary men of the time exploit with critical profit for everybody. But although the stresses register increasing strains, they do not bring a breakdown in communication; rather they transmit motion from one part of the spreading field to another, and in a loose but reasonable harmony of opposites.

There were men who were sensitive and responsive to the oppositions. In a theological tradition that goes back at least to Tertullian, and possibly to Heraclitus, there was a habit of faith that expressed itself in various methods of harmonizing opposites. Sometimes this method consisted in an ardent ascent above reason to mystical vision or mysterious dogma, as in the case of Tertullian, who said, "I believe because it is impossible." Others like Heraclitus argued that all contradictions are composed as the contradiction between being and nonbeing is combined in change, or as good and evil meet in heroic tragedy. This theme had been picked up early in the Renaissance by Nicholas of Cusa in a book, *The Learned Ignorance,* which used the paradoxes of the mathematical infinite to expound the doctrines of the Incarnation and the Trinity. It is now known that the paintings of Raphael and Michelangelo used discordant perspectives as well as Biblical drama to heal these scandals of speculation. One climbs to wisdom on a ladder of paradoxes.

But one should not mistake this obbligato of dialectic, which has a continuous history as a hatred of or retreat from reason. Dialecticians may enjoy their art as other artisans do, but they do not practice without provocation or without the stimulation of the unavoidable failures of reason. Dialectic of this sort is the practice of reason to defend and heal reason.

Another method by which strains and stresses are brought into dynamic equilibriums is suggested in the music of Palestrina. Up to his time it is said there was no explicit recognition of harmony as distinct from the patterns of melody. The Greek modes and the modes of plain song were understood as the matrices for single-voiced compositions. If more than one voice followed different melodies at the same time, they were simply supposed not to interfere by cacophony. One of the reasons for this single interpretation for music may have been a fear of those points in the musical continuum where the modes which composed it overlapped one another and were in conflict. Melodies could avoid these with ingenuity in composition, but chords with melodies might run into serious paradoxes. Palestrina, learned and skilled in all the modes, and using them in combination to invent others, introduced many voices, sometimes many more than any modern orchestral score prescribes, and kept them all singing in rounds, motets, fugues, and other, unnamed combinations. He made the musical objects from which our classical and modern music are derived, but perhaps more important, the musical objects which have become the subject matter for modern theories of harmony and melody. He forced music out of the standard forms in which the paradoxes threatened to destroy it. The fearsome difficulties are at least postponed.

Reading Kepler is like listening to Palestrina. He has many voices singing together, voices of poetic allegory, both theological and psychological, voices of number, figure, ratio, and the new analysis. The effect is intoxicating, hypnotic, and otherworldly, but wherever the reader stops to calculate or to check an analogy, he finds harmony—mathematical, factual, and poetic—and validity within the limits of the observation and instrumentation of the time. The paradoxes that are being avoided are all connected with the literal interpretation of the dogma that the heavenly bodies must move regularly in circular orbits. Both Ptolemy and Copernicus had used eccentric circles to keep their accounts straight, but they had thought of no other figures. Kepler preserved the regularity that his mathematics allowed and required but let the circles transform themselves into their near neighbors, ellipses.

Kepler accepts as his most immediate teachers Tycho Brahe, who had developed a system intermediate between Ptolemy and Copernicus to go with his exhaustive tables of observations, and Gilbert, who supplied the material for the best guess about the motive power for the celestial machine—magnetism. But he was enormously learned in all the ancient lore of astronomy, mathematics, music, and poetry. The creative fire in his mind at first seems to be religious, but the total impression is that of a poet who ascends, as he says, with the Muses to Apollo. The temptation is to compare him with Dante, but then it is obvious that he would have pleased Plotinus more.

One of his startling introductory images is derived from the scholastic doctrines in which things in the world are so arranged and constructed that they show traces of the Trinity. The scholastic accounts show long trains of these traces in the human mind, in nature, and in the varieties of human artifice. Kepler invents a new trace; the sun is the Father, the surface of the sphere of the fixed stars is the Son, and the space in between the sun and the sphere is the place of the Holy Spirit, within which the planets float in an ethereal fluid. The modern reader may suppose this is a somewhat blasphemous whimsy, picked up idly by a witty man, but there is evidence from its use that it is a part of an organic structure, demanded implicitly by the method and the system. The image, at least, of God is needed in his hypotheses.

Then he goes on with the properties of those pervasive members of the great world, light, heat, movement, and harmony of movement. Each of these properties emanates from the sun, and is reflected back to it by the sphere of the fixed stars. Each of these properties is also analogous with a faculty of the human soul: light to the senses, heat to the vegetative faculty, movement to the animal, and harmony to the rational faculties. Here again the modern reader's attention is thrown off by what seems to be a piece of gratuitous fantasy characteristic of didactic fables for children, sugar coating for dogma, or the obfuscation of the occult; but the skeptical defenses are broken down by the fullness of conception, the adequacy of the mathematical analysis, and the faithfulness to the tables of observation inherited from Tycho Brahe. The wholeness of conception is poetic in its style, but wholly responsible to principle and fact in its hypothetical development. There is practically nothing that has since happened in science that is not foreshadowed in Kepler, and that could not be assimilated to his system.

It is also true that most of the ancient mathematics, physics, astronomy, physiology, biology, and even psychology plays some role in the Keplerian system. In an obviously triumphant spirit the relative angular speeds of the planets as seen from the sun are shown to conform with the numbers that express the musical intervals, so that it is literally true to say that early modern music is the presentation of the harmony of the spheres to human hearing. Also, the relative distances of the planets from the sun are shown to conform with the radii of spheres that are alternately inscribed and circumscribed in the series of the five regular solids. Finally, the orbits of the planets are shown to conform to the ellipses that have become canonical for the modern solar system. In all this there is a completely knowing recognition that hypotheses are the main subject matter of his constructions, so that a subsequent revision of data or the discovery of new planets, such as has taken place, would have incited new construction rather than threatened precious or dogmatic theory.

It is said that this formulation was done over a long period of years with sustained excitement and prodigious labor. Tycho Brahe's life work had been the complete review and revision of all astronomical observations and records. Most of this had been done under the guidance of the Ptolemaic hypotheses, but the corrections and additions which Brahe's own observations had made led him to the hypothesis that five of the seven planets moved around the sun as fixed center, but that the earth did not move. The two worked together for a time, and on Brahe's death Kepler continued. Kepler's conclusions concerning the arrangement and working of the solar system are well known—the relative spacing of the planets, the elliptical orbits that replaced the Copernican circles, and finally the regularly irregular velocities of the several planets according to the law that the radius vector, the line from the sun to the planet, sweeps out equal areas of the ellipse in equal times.

This last law alone not only gives Kepler his heroic place in the making of astronomy, but it marks the complete conquest of the new analytic method. The whole astronomical enterprise had outgrown the cumbersome devices, both instruments and mathematical conceptions, not only of the Ptolemaic system, but of the Copernican as well. Kepler, his mind packed with all the knowledge that these systems had accumulated, and supported by the courage of his own imagination, made the bold intuitive leap into the realm of the infinitude of possible figures. That he should fasten upon the ellipse is not surprising; it is merely a flattened circle; but that he should be able with his arsenal of ratios and proportions to trace the order and connections in the mountain of observational data with which he was faced and come out with a proportion of times and areas—this is enough to persuade one of angelic revelation, and perhaps angelic guidance for this whole period of science, as well as for the planets in their courses.[19]

One is sure in reading Kepler that one is not following more than a small fraction of the insights and connections, and one is almost sure that Kepler himself does not comprehend the depth of his wit. To be sure, there are many threads that lead to the secret, some known and some that it is likely will be better known when the intellectual history is better done. For instance, the ancient solution to the problem of the irrational number and magnitude in terms of "taking the irrational in square" areas gives us a glimpse into the lesser-known background of Kepler's thought.[20]

The strains and stresses that show between the paired heroes of this century, the implicit enmity between love and the machine, the antipathy between the mathematician and the physicist or the mechanist, reach the resolution that they seem to demand in Kepler. He sees the motions and paths of the planets within the ocean of the Holy Spirit as arranged by intelligence, in mathematically lovely patterns for the delectation of the contemplative creature, man. But then he does not rest on his speculative oars as teleologists are supposed to do. He turns to the problem of the mechanical motive power and articulation that the solar system presents. Instead of using the work of Galileo as Newton later did, he uses Gilbert. Out of his conception of the polar attractions and repulsions of the magnet, Kepler fashions a celestial machine which approximates the operations of the modern electric induction

[19] As Buchanan is implying, Kepler had difficulty enough becoming convinced of the rightness of this area rule, but it is worth noting that it is a strict consequence of a prior Keplerian assumption, namely that the planet is being pushed about by a "motive virtue" issuing from the sun, the strength of this virtue varying inversely with distance from the sun. It follows that the component of the planet's velocity at right angles to the radius vector is inversely as the distance from the sun, and this is equivalent to the area rule.

[20] Kepler uses the hierarchy of rational and irrational magnitudes in Euclid's *Elements,* Book X, to construct a hierarchy of harmonies, exemplified both in musical consonances and in planetary "aspects"—the angles between planets as seen from the earth that are effective in exciting the earth-soul into extraordinary activity.

motor, with the sun as an internal rotating field and the planets units of a vast and complicated external rotating field. It is not difficult to make a model of such a motor with two bar magnets. The secret of the conception is the half-imaginary and half-perceptual image of the lines of force in a magnetic field as demonstrated by Faraday with iron filings. The sun and the planets have "threads" that push and pull in a ghostly imitation of clockwork. There is an inverse proportionality of distance and intensity of the forces that suggests the Newtonian inverse square law and Faraday's lines of force.

In the present fragmentary state of the history of science one cannot read Kepler without a poignant sense of the fateful ironies in intellectual history. Kepler seems to belong with Plato and Dante to that small society of minds lit with a poetic fire whose light reaches through an intensely human medium to the darkest corners of the universe. In a superficial way they are lonely persons, half-exiled and walled in, but their prisons are very high towers, in the tops of which they live and watch. They are supported as much as they are banished by their immediate intellectual communities. To use a phrase of Arnold Toynbee, they withdraw and return, but the etherealized product of their lonely work is not received in an adequate human receptacle. Parts of it are received and put to work usually in quite separate and uncommunicative vessels; the form of the whole, to quote Plato, is "laid up in heaven." One wonders what would have been the consequences if the Platonic Academy had been established in Syracuse instead of in disintegrating Athens; what would have happened if Dante's Florence had flowered with laboratories as well as studios; what would have crowned the century of genius if Kepler's Prague had become the home of the European academies rather than London and Paris. These romantic ironies of history are probably more expressive of the pathos of the mid-twentieth century than they are of any steady and just view of the intrinsic merits of the minds of these men or of their real influence on history. It is probably true, however, that the brilliance of their visions dazzled and blinded as well as inspired the minds of their contemporaries.

At any rate, the heavenly fire of Kepler's mind seems to have been brought to earth not by a Prometheus, but by a Hephaestus. There is a feeling that all these heroes were moving upward and downward in the ethereal medium between earth and the heavens of the fixed stars. But there were indications of what would happen in the objects Galileo studied—the public works of Venice, glass for the telescopes, levers, pulleys, beams, inclined planes, and unrepaired leaning towers for the theory of falling bodies. These are the affairs of the workshop, and it was from these that Huygens and Newton took their cues.

Huygens and Newton both started by taking the telescope apart, and the problems in optics that they found in the lens occupied each for long periods. The lens of a telescope performs its function by refracting or bending the rays of light; a lens is in fact a complex prism, one of whose properties is to split white light into its component colors. As a lens does this, it produces what are called chromatic aberrations, as well as magnification; the lens does not exactly magnify the errors, but it produces faulty images. Newton worked on the problem, hoping to find a cure for the colors, and finally proved to his own satisfaction that there was no cure for the lens. He thereupon invented the reflecting telescope which avoided the troubles of refraction. In the course of the work he developed a general theory of optics and of color which has been classic since his time. Incidentally, in his theory of light, he assumed that the light ray consisted of little bodies, or corpuscles, moving a hundred ninety thousand miles per second, and having peculiar shapes that got tangled in the mediums through which they passed. In place of the "fits of the corpuscles" Huygens developed the analogy between the propagation of waves in water, of sound in the atmosphere, and of light in the ether. This is a bold theory, which brings many sharp distinctions of precise observation and many subtleties of mathematical construction into an accord "with the principles accepted in the Philosophy of the present day"—by which phrase Huygens refers to the new Theory in which he found "not only the ellipses, hyperbolas, and other curves which Mr. Descartes has ingeniously invented for this purpose; but also those which the surface of the glass lens ought to possess when its other surface is given as spherical or plane, or of any other figure that may be." Although Newton and Huygens show a deep disagreement in their theories of light, a disagreement that persists even to now with all the new data and variations in explanation that have accumulated, they agree in their observations where their selections of phenomena coincide, and in the formal geometric descriptions of reflection and refraction. Between them they also report phenomena that have later been called diffraction, interference, and polarization of light.

But in spite of the elegance and clarity of their discourses, which are mathematical and mechanical in theme, there is a new atmosphere in their enterprises. Where it seems that Galileo was observing and generalizing from industrial and engineering operations, Huygens and Newton have set up a new kind of shop for themselves; they have turned the workshop into a laboratory. It is known that many monasteries and many learned scholastics discovered a special kind of joy and efficiency in combining the activities of the garden, the workshop, the observatory, and the library, with the result that the Church is credited with having harbored and perpetuated magic and superstition in the form of astrology, alchemy, and necromancy, as well as the liberal and mechanical arts. It is often forgotten that these activities continued, but it is not surprising to find that they had secular imitations. Huygens and Newton not only had their laboratories; they reported their findings to Royal Societies and Academies. Dukes and earls had often subsidized the printing of reports from learned persons; in the seventeenth century kings were subsidizing the pursuit of science, and establishing societies of scientists who engaged in the political arts of deliberation and advised the appropriation of funds as the incipient parliaments were doing. The scientist had a parliament of peers to judge and support, and therefore to control his efforts. It is not to be forgotten that science and its connection with industry and commerce have had an increasing interest for rulers and parliaments in the last three hundred years. It may be that the industrial revolution started in the monasteries, but the pattern of science, industry, and government that gives its more commonplace and disturbing meaning at present was not formed until the seventeenth century. With Huygens and Newton, although they seem to be unaware of it, its fateful pattern is fixed.

The Baconian Revolution

It was inevitable that the industrial revolution should bring with it a revolution in scientific method.[21] This revolution had been announced a half-century before the births of Huygens and Newton by a man who was primarily a jurist and a statesman for whom science was apparently only a hobby, Francis Bacon. As a very young man he had been impressed by the political and practical irresponsibility of the speculative mind. Toward the end of a rather devious and colorful life of political ups and downs he returned to his youthful concern with this problem, and published several books announcing a new epoch of history in which the good of humanity would be served by science.

The major part of these books is devoted to a new arrangement of human knowledge based on his own division of it into three parts: memory, imagination, and reason. On rightly appraising these parts it would be clear that there must be a new instauration or revival, and an

[21] It should be emphasized here that Buchanan is taking "industrial revolution" in a broad sense that can cover the introduction of the spring-pole lathe in the thirteenth century, the suction pump for mines in the fifteenth century, or the overshot water wheel in the sixteenth century, as well as the steam engine and spinning jenny in the eighteenth century.

advancement of knowledge based on a method of inquiry, a new organum or instrument. The new world that would result was called the New Atlantis. A part of one of these books is called the *Novum Organum*. It warns the user of the new instrument against the Idols of the Tribe, arguments from tradition and authority; the Idols of the Cave, prejudices due to individual temperament and habit; Idols of the Marketplace, errors due to the imperfection and misuse of words and symbols of communication; and Idols of the Theater, dominating systems of thought that amount to arbitrary dogma. These are not, as often supposed, condemnations of avoidable fallacies; they are warnings against the politically naïve use of necessary aids to investigation. Even so, their concern with human welfare tends to sweep the field clean of any historic or traditional method of science. Perhaps they were particularly directed against Aristotle's influence, but they also rule out any concerted effort to use the new analytic method that dominated the time. This seems to argue that Bacon must have been a very ignorant man, but on second thought it is quite probable that he knew what he was saying, and that he was right if we take him for what he was, a judge of the social, economic and political influences of science. There is a serious meaning in the oft-quoted witticism that he discussed science like a Lord Chancellor. There is no doubt today that scientific knowledge is power—economic, social, and political.

The method he proposes should then be judged in terms of ends and aims of science that had not been before admitted, ends that seem more relevant to the black arts of magic and mystery than to the liberal arts and sciences. We have to admit the legitimacy and political astuteness of the point of view, if we are not to condemn the last three centuries of the scientific and industrial revolution.

The rules of the method set up a kind of factual bookkeeping. Having washed the mind of all principles and hypotheses, which may carry with them unknown institutional implications, the inquirer into nature makes an exhaustive Table of Presentation of Instances to the Understanding, in which there is first a Table of Essence and Presence, "instances which agree in the same nature, though in substances the most unlike"; a Table of Deviation or of Absence in Proximity, "instances in which the given nature is wanting; because the Form as stated above ought no less to be absent when the given nature is absent, than present when it is present"; a Table of Degrees or Comparison, "instances in which the given nature is found in different degrees, more or less." Although the exposition of the method is provokingly brief, and to the operating scientist, as well as to the critical philosopher, inadequate and misleading, the illustration given to aid the reader is spectacular and prophetic. Bacon actually draws up the table of instances which rhetorically present an impressive mass of evidence that heat is identical with motion. This is a canny anticipation of the kinetic theory of heat. We can also now remind ourselves of the scientific method of Thomas Edison, if we need more circumstantial evidence on the validity of the method. It is more than likely that Bacon thinks that science is a subtopic in procedural law, the law of evidence. In some jurisdictions in our contemporary world, not all of them totalitarian, organized science has become a subtopic in politics.

Actually, there have been minds better instructed in scientific procedures than Bacon's that have refined this method and made it the account of scientific progress in the last three hundred years. It is extreme in its drastic exclusion of hypothesis and principle, where the more refined versions have merely subordinated theory to the establishment, collection, and arrangement of facts. Besides the new recognition of the part played by science in other human affairs, the *Novum Organum,* as a revolutionary manifesto, is justly reporting the increasing weight of the new data that had accumulated in the practices of older methods that were aimed at the establishment of principles and the discovery of hypotheses. Many of the facts that Bacon adduces in support of his kinetic theory of heat are by-products of the observations and researches made under the rules of the older methods, and they would not have been noticed if the old rules had not been taken for granted. They were discerned and certified by hypothesis and principle, and they therefore carry the significance and force of principle and hypothesis implicitly. In effect, Bacon is asking for an inverse operation to intuitive induction: restore the principle to the facts. It will be noted that he is aware of the presence and absence of "forms."

This interpretation of Bacon may help to explain the troublesome statement of Newton, a statement made twice, once in connection with his work on optics, and once in connection with the theory of gravity, that he made no hypotheses. This can hardly mean that he used other men's hypotheses and made no new ones of his own, as his other oft-quoted remark that he had stood on the shoulders of giants, and thus seen a little farther than they had, suggests. It rather means that he was seeing mathematics separated and purified of its speculative and poetic associations, and at the same time, implicit, implicated, and applied to the observations and operations of the laboratory. It would seem that the new analytic method had become a matter of fact.

Newton

It is not surprising that this should have happened. By Newton's time the new method had very wide acceptance and the certification of many successful applications. Then too, Newton had a piercing though narrow brilliance of mind. He had read Euclid's and Descartes' geometries with excitement when he was quite young, and the two together with the teaching of Wallis, the author of a book on the *Arithmetic of Infinites,* had stirred his mind to see and to formulate the Binomial Theorem from which as origin and model the most powerful part of modern mathematical analysis is derived. The pious inclusion of the theorem in most elementary texts in algebra today leaves it a puzzle, and an easy puzzle at that, for most school boys. This is because its generalization is not mentioned. The Binomial Theorem is a simple elegant case of a general principle that any algebraic function contains an internal structure that can be represented by a series, as a binomial raised to any power can be expanded into a series. This principle and the algebraic operations that go with it are implicit in the development of algebra from the Greek calculus of ratios and proportions—as to be sure they are implicit in the fact that a number is the result of counting—but the form that the principle took in Newton's mind amounts to a new view of all mathematics. For other minds—for instance for Leibniz—it represented a new mansion in the heaven of mathematical forms and a new universal language in which reason might master the world, but for Newton it was a new art for combining observations, and he took it into the laboratory.

By Newton's time there had occurred another shift in the weights of scientific concern. Most of the crucial discoveries and formulations had involved expressions for the quantification of time. This had always been the case in astronomy and had not been missing in the physics of local motion, but the new style in Kepler's astronomy and in Galileo's account of falling bodies had made time an essential constituent of the application of mathematics to nature. Stemming from Plato there was a tradition that time was the moving shadow of eternity, and from Aristotle there arose controversy whether time was a measure of motion or motion a measure of time. But the work of Galileo and Kepler which could be taken to show that local motion was completely analyzable in terms of space and time suggested that time was not merely a technique in measurement but a constituent formal part of mathematics itself, the as yet undiscovered part of the modern art of analysis. Time was a numbered series, Aristotle had said. Now Newton in effect said that it was the principle of all series, and that all variables in algebra are flowing quantities. There are simpler ways of saying this, as Kant later showed: space is the external form of our perceptions; time is the internal form of our perceptions. From the former arises the science of geometry, from the latter arithmetic. Since analytic geometry is the combination of these two sciences, time must be part of mathematical analysis.

One result of this assimilation of time to mathematics is Newton's theory of fluxions, or flowing quantities. If the terms of a ratio are flowing quantities, or represent series of numbers, as they do, then there is a velocity of the flow, and a possible acceleration or change of velocity of the flow, and these may be represented by still other ratios between quantity and time. These ratios can be represented by variables or more explicitly by the series that they generate. This without the machinery of symbols which is necessary for its use is the heart of the theory of fluxions, or as it is called after Leibniz's formulation of it, the differential and integral calculus.

As time was thus being built into mathematics by Newton, his contemporary, Huygens, was building it into machinery. He did this by taking over Galileo's work on the pendulum whose periodic movements had to be attached to a system of cogwheels which transmitted motion from a pulley and weight to the hands of the clock. This was accomplished by the small vibrating mechanism, the escapement, which alternately stopped and released a cogwheel meshed with the main gears. It is interesting to note that the certification of the regularity of the periodic motion of the pendulum was checked first by the human pulse and then by the water clock as it was used in Galileo's experiments with the inclined plane. By this circuitous route the small portable machines now so familiar to us were made to imitate the periodic motions of the stars. With the mechanical clock the refined measures of time came into the laboratory and the observatory where they met the equations of motion and the accompanying series which Newton had prepared for them. With the mechanical clock to measure time, and with the theory of fluxions to assimilate the place and time data inherited from the Ptolemaic observers and corrected by Tycho Brahe and Kepler, Newton proceeded to build a bridge of intelligibility between Galileo's laws of falling bodies and Kepler's orbits of the planets.

The popular version of this intellectual deed is the story of Newton's contemplation of the identity of law exemplified in the falling apple and the motions of the moon. The full report of it is in *Principia Mathematica Philosophiae Naturalis,* or *Mathematical Principles of Natural Philosophy,* which is a curious piece of writing forced from the hand of Newton by his importunate friends in the Royal Society. It is written in at least three languages or dialects of science; the language of Greek mathematics involving the traditional ratios, proportions, and figures; a language alluding somewhat darkly to the theory of fluxions and algebraic analysis, and a new language of matter and force. The reader has very little warning of the semantic tangle through which he has to pass. Newton constructed the tangle because he supposed that the theory of fluxions would not be understood and therefore would give rise to empty controversy; as a matter of fact his subterfuge prevented the book from being read and understood on the Continent for fifty years, until Voltaire wrote a popular version of it. Even in England it had to be rewritten and commented on by several people before it delivered its teaching. It is a riddle for the reader of the great books of Western science, a kind of perpetual final examination for one who is reading and rereading the books that surround and support it. Few people, including mathematicians and physicists can disentangle the thread of the argument throughout the book, especially in the second book where several episodic problems in physics are brought into accord with the main theory. But by reversing Newton's admitted device of exposition, that is, translating the terms of the language of ratio and proportion into the language of analysis and fluxions, and translating the language of matter and force into mathematics, the tangle can be unravelled. For this process, as well as for understanding the methods of the modern period, the key seems to be the mathematization of time.

For instance, Galileo had formulated the first law of motion, that a body continues in a state of rest or of uniform motion in a straight line unless it is impressed by an external force.[22] This means that if a body suffers acceleration by changing its velocity or its direction, it is being influenced by an external force. This is a generalization from the change of velocity found in the motion of falling bodies. Similarly, the acceleration, or the change in direction and velocity, in the parabolic flight of a projectile is due to the influence of an external force, the pull of the earth. This is partly expressed and partly suggested in the second law of motion, which says that the force impressed is proportional to the acceleration or change in velocity and direction of a moving body. Applied to the planets moving in their Keplerian orbits, the accelerations, decelerations, and elliptical paths show mutual attractive forces at work, measured, as Newton says, by the degrees of their ac-

[22] The first law of motion in the form here stated is attributed by Newton to Galileo, but it does not appear to be the case that Galileo ever thought of the continuing uniform motion as rectilinear; rather, he described it as horizontal or parallel to the horizon, and so by implication circular. In the case of terrestrial bodies, Galileo never abstracts corporeality from weight; hence his physics remains a physics that is essentially bound to the vertical and horizontal directions. Even in the case of celestial bodies, he fails to envisage the operation of a rectilinear inertia.

celerations. The formula of these accelerations or forces says that they are directly proportional to the masses of the respective bodies and inversely proportional to the square of their distances. This is the famous law of gravitation applicable to all bodies; it explains weight as the acceleration of bodies toward the earth, and it explains the regular irregular movements of the planets in their elliptical orbits. It also explains the complicated relation between the tides of the ocean and the vagaries of the moon. Its promise of universal application wherever there is motion in the universe, and the expectation that all phenomena would somehow fall under its rule, open and chart an epoch in modern science.

Like most revolutionary formulae the "Newtonian laws of motion" cast a new light on the past. Descartes and that company of heroes who shared his efforts in the new analytic art would have been surprised at the outcome of its application. Where for them the method somehow shifted interest away from the metaphysical heights of the One and the Deity, it opened a wide and spacious (if not spatial) realm a little lower than Kepler's planetary souls. It is true that the floor of this realm was extension and motion, but its ceiling was unlimited; there was room but no need for Kepler's celestial motors and souls to produce and to understand the music of the equations if not the harmony of the spheres. The effect of the Newtonian synthesis was to bring all these hypotheses down to earth, or rather to find them immanent in the data. The ancient doctrine that natural motion was innately circular and the motion of perfect bodies, while rectilinear motion was violent and imperfect because caused by external causes, was broken in the first law of motion. Newton finished its destruction by reducing the planetary motions completely to the continuous resolution of the rectilinear parallelogram of forces. Furthermore, he restricted the properties of bodies for the methodical notice of natural philosophers to their space and time coordinates, to their simple location in space and in time, as A. N. Whitehead has said.

If one looks at the law of gravitation and the laws of motion together with the definitions at the beginning of the first book of the *Principia,* there seems to be an exception to this account of the reduction. The notion of mass as the measured quantity of matter in a body seems to resist the reduction, but if the notion is properly distinguished from weight, and its derivation from density and volume examined, it appears to reduce itself to something like prime or signate matter, an almost pure potentiality for motion, or to a fitness for entering the equations of motion. In the controversy between the Newtonian and the Leibnizian formulation of the calculus, the British partisans of Newton accused the Continental partisans of the Leibnizian infinitesimal of dealing with the "ghosts of departed magnitudes"; it might be said that Newton in his treatment of "mass" as numbered matter is dealing with the "ghosts of banished bodies." Actually what had happened was that bodies in the gravitational system have to be conceived as points in a figure, as the weight of a pendulum has to be conceived as concentrated at a point. But the points have to be numbered or lettered in the analytic system to enable them to satisfy the equations, and these numbers are merely terms in the ratios of velocity and acceleration. The mysterious absoluteness of the terms for mass, the kind of absoluteness that Newton saw in absolute space and absolute time, showed the strain of the whole revolution in thought. Mass is a kind of umbilical cord reaching back to being from the analytic systems; it has taken the midwifery of Einstein to free the offspring from any trace of its parent. Mass has now been dissolved in the equations of energy, and in the fearsome fog of the atomic cloud.

Newton in effect traced and tied off the last knots in the net of ratios that surrounds and holds the modern world. Fundamentally and simply this net reflects an infinite set of spatial and temporal relations, but these would not be seen or in a sense be there, were it not for the analytic method and its rules. Ignoring the ghostly reminder of "mass," and some of the other puzzling "constants" of physics, the ratios of time and space have lost any connection they once had with the allegories that once pierced the ceiling. Instead the modes of their signification not only reflect the arrangements in fact and data; they also map and direct the observation and experimentation that together constitute the going enterprise of science since Newton's time. They supply the geography and the schedule for the collection and arranging of the specimens. The modern museum, the catalogues of minerals, plants, animals, and stars, and the modern encyclopedia store the constantly accumulating results.

There are two other results that concern method. One is the leading role that time plays in all of our thinking. This takes its rise in the Newtonian equations and spreads to all fields, so that natural history has given way to theories of evolution not only for plants and animals, but for everything from elements—suggesting alchemy—to human institutions—suggesting astrology. There are few things in the celestial hierarchy that do not appear in the annals of time past or future.

The other result is the companion of this temporalization. Duration, which is the name for what the clock measures, is a felt dimension or direction in human experience. St. Augustine had said of it: "If no one asks me, I know; if I want to explain it to a questioner, I do not know."[23] With the Newtonian system, we have come to be able to

[23] *The Confessions of St. Augustine,* tr. F. J. Sheed (London and New York: Sheed and Ward, 1944), Book XI, Section xiv, p. 217.

answer Augustine's question: Time is the order of experience, the hand of Providence arranging the items in experience; as clocks tell us when to look and act, so felt time supplies us what to look for and what to do about it. It even seems to supply us with life. Experience is therefore the source from which we get the stuff that goes into our scientific understanding. In obeying the mechanical clock we include ourselves in the process of scientific investigation. We verify hypotheses almost by living. But when this is not enough, we explore and experiment, we build machines that embody the laws of motion and we turn them into instruments which discover more laws of motion. As the Pythian maiden once sat over the cleft in the rock at Delphi and articulated in numbered verses the wisdom of Apollo, the god of reason, so now Experience navigates the waves of chaos shooting the sun and stars and establishing its daily position according to the numbered intervals of the chronometer.

Empiricism and Rationalism

It took a hundred years for the scientific world to assimilate the calculus and the law of gravitation, not merely because gravitation as a system had to displace the widely accepted Cartesian system on the Continent, and the Newtonian form of the calculus had to come to terms with the Leibnizian method of infinitesimals, but also because Newton's—and in general the British—understanding of the method and system carried with it the new emphasis and weight of experience and experiment in the scientific enterprise. Beginning with Locke, a contemporary of Newton, and gaining virulent power in Berkeley and Hume, British philosophers attacked and reattacked what they saw as the established scholasticism and rationalism of the Continental system makers.

Many of the special objects of criticism appear now as straw men, doctrines that the Britishers formulated for the first time, which were never dreamed of on the Continent, or if held at all by Continental philosophers were not of central importance. As an example, Locke refuted at length the doctrine of innate ideas, the doctrine that men are born with self-evident propositions in their minds from which they then deduce the truths of science and metaphysics. This doctrine is a parody on Descartes's method of intuition and deduction, which presupposes the relevant powers of the mind, the successful use of which requires a discipline and a method. The simple, clear, and distinct ideas are no more innate than Locke's sensations are; both are to be attained by the operation of powers of the mind in experience. On the other hand, Descartes and many other followers of the analytic method did accept and trust the validity of abstract ideas, and this is the center around which the attacks of the empiricists converged.

They attack first the ideas that were in the Aristotelian tradition called genera and species, the common properties of more than one individual, such as the idea of horse, or the idea of house. Locke shows that these are arrived at not by abstraction but by the addition of simple sense data that in combination characterize familiar perceptual objects and determine classes or collections of individuals. Berkeley shows how these ideas arise by the subtraction of simple ideas from the complex ideas that characterize the individuals. He shows that the idea of dog treated in this subtractive manner loses all characteristics and approaches zero as its content. Hume finds it impossible to see any essential connection between simple ideas at all, so that even perceptual objects are atomized, and abstractions become merely the habits by which we view and arrange the items in our kaleidoscopic experience.

The kind of abstraction that mathematics deals with gets more respect and solicitude. Locke, being close to Newton, finds the ideas of size, shape, motion, and even impenetrability primary qualities of things, whereas, following Galileo, he thinks that color, smell, and the other ideas associated with the senses are secondary, actually effects in the mind alone and attributed only by inference to the object. Berkeley with incisive logic shows that this distinction is not justified; primary qualities also exist only in the perceiving mind. On the other hand mathematics, which deals with these primary qualities, is the science of necessary connection, and there is certainty in the discursive reasoning that is based on them, a certainty not of existence, but of deduction. There is no evidence that mathematics applies to anything outside the mind.

The effect of this radical empiricism on the uninitiated reader can be very devastating and at the same time very releasing. On the other hand it is not as complete a scepticism as it at first seems. The criticism is made in behalf of a new method, the method of combining refined observations to implement the program of Francis Bacon, to make science effective in the practical control of human affairs. Locke was a physician and a political scientist; Berkeley was a clergyman and a bishop who came to America to found an institution of useful learning at New Haven; and Hume was a historian. They are concerned about science as a branch of practical reason, and they are attacking its apparently single-minded devotion to speculative truth.

Spinoza and Leibniz, the two Continental philosophers who followed

Descartes in espousing the new analytic method, came into the controversy with metaphysical elaboration of the analytic style of reasoning. Neither of them is a good dialectician, so their theological metaphysics is merely a weak imitation of the thought of the scholastics, involving risks of naiveté in metaphysics and heresy with respect to religion. They do not go in for allegorical development of their hypotheses in the style of Kepler. Instead they generalize and etherealize the mathematical constructions until they become attributes or creative ideas in the mind of God. For Spinoza, Descartes's substances of extension and thought, which are really purifications of the geometry and algebra respectively of analytic geometry, are made parallel attributes of the one substance, God. For Leibniz, the mathematical functions represented by equations expand to become independent closed systems, possible worlds in God's mind, and the infinitesimals become monads or the spiritual atoms that constitute these worlds. Both tend to view the world *sub specie aeternitatis* instead of *sub specie temporis,* as is well illustrated even in the Leibnizian formulation of the calculus free of all literal reference to time. As a corollary of this upward thrust of abstraction from the analytic machinery, both Spinoza and Leibniz ignore or depreciate the role of experiment in scientific procedures, and they think of sensation, memory and imagination as confused knowledge, at best only suggestive aids in the acquisition of clear and distinct ideas. It is interesting to note that Leibniz, as if to marshal the social and political forces on his side of the argument, helped to found the Berlin Academy of Science.

The resolution of the controversy that continued for two hundred years over the English Channel concerning the interpretation of the analytic method as it focused itself in the calculus has several minor ironies. The theory and practice of the fluxional analysis finally gave way even in British universities to the Leibnizian use of derivatives and integrals, but aside from the mere mathematical elegance of Leibniz's symbolism, the reason for its victory was the greater facility it afforded for application to the empirical findings of the laboratory. This might have pleased Leibniz as an unexpected utility for his invention, but he would not have agreed to it as the interpretation of its much discussed meaning. At the same time, the partial resolution of the controversy concerning scientific method in general, of which the calculus was a part, is to be found in the *Critique of Pure Reason,* in which the main theme is the praise of time and imagination as the essential themes in scientific method.

Kant's Criticism

KANT says that he was roused from his dogmatic slumbers in a rather second-rate combination of scholastic logic and Leibnizian analysis by the skeptical writings of Hume. Kant shared a concern for moral philosophy with the British empiricists, and an enthusiasm and facility in mathematics with his Continental companions, both of which amounted to awe when he viewed "the starry heavens above and the moral law within." He also shared the growing skepticism concerning dialectical and allegorical explorations of dogmatic and natural theology. He could see the advantage of shifting the burden of certification from the paradoxes of speculation to the data of experience. He found the device for making this shift in mathematics itself.

He begins by showing that the operation of the intuitive powers by which the stuff of science is introduced into a system of concepts is the initial and unavoidable condition for the human mind having experience at all. These powers consist of formal patterns of space and time in which all sensations are framed; the data of experience are given here and there, now and then. Such data are the product of the productive imagination.

But it is also true that each concept or universal in the mind gives rise to spatial and temporal series. Each concept we have contains within it a formal pattern or scheme of wholes and parts and of items in a series. The reordering of the stuff of experience already given in certain temporal and spatial patterns into the conceptual schemes is the work of the reproductive imagination. Parts of these schemes are lifted from the matrix of experience unchanged in their order; this is memory. Parts of them are rearranged by construction to fit conceptual forms. All valid judgments—quantitative, qualitative, causal, hypothetical necessary, or actual—assent to and assert the correspondence and fitness of the items of experience with the conceptual frameworks. Another way of seeing this is by noting that each concept is a rule for ordering or reordering the manifold of sensations that experience presents. Certain concepts, like the ideas that Plato could never find embodied, are rules for finding and ordering the concepts which are to be embodied; these are ideas of reason as the embodied concepts are concepts of the understanding.

It is important to see where this neat theory comes from, for it is by no means all Kant's invention. First, it is a further analysis of the intuitive induction, in which a single observation allows an inference. For Kant the intuition is our immediate grasp of space and time in perception; the inference can validly follow because the intuition is an item or an interval in the series of conceptual schemes. The fact has found its role in an intelligible plot. The serial order comes from the work of Nicolas Oresme who diagrammed the variations in all qualities, so that they form a series. The identification of serial order and time is from Newton, as he saw it in the expansion of functions into series. It thus appears that the analytic method in its increasingly systematic applications during the three previous centuries had been preparing a mathematical receptacle for time, and therefore for its children, the manifold items that flow in the stream of experience. Analytic geometry had adjusted the two mathematical mediums of space and number, geometry and arithmetic, so that together they could be focused on any point or interval in space; this established extension or space and time as systematized in analytic geometry, as one essential condition for the application of scientific method. With the invention of the calculus time was built into the mathematical instruments of investigation and reasoning.

The first part of the *Critique of Pure Reason,* by certifying that the items of experience are inevitably given in a framework of time and space, confers upon experience itself as a continuous whole the character and function of a mathematical discipline. Experience is intrinsically and unavoidably mathematical, and through it all observation and experimentation are automatically mathematical. This not only resolves the issues between the empiricist and the rationalist; it is a new charter for the scientific enterprise. Where intuitive induction was once the narrow strait, found with difficulty and only by heroes and geniuses, for the introduction of fact to the heaven of theory, it was now assured that all phenomena contain the divine spark of intelligibility and array themselves in constellations of understanding.

The reader might suppose that this would give a new importance to the great hypotheses of the Renaissance, and in one respect this is true. The nineteenth century saw a development of mathematics scarcely equalled in comprehension and complication in all previous history, and going with this a skill and subtlety of application never before approached. But at the same time the status of mathematics as a standard and model of certain knowledge suffered a progressive decline. It became a set of rules, or an organon, for the refinement of experience in measurement and experiment, and for the establishment of fact. The clear and distinct ideas that almost alone had been trusted to generate and maintain the great hypotheses came under suspicion as bright fictions and seducers into vain speculation.

Kant records this change in the temper of scientific investigation in the well-known aphorism: Intuitions without concepts are blind: thoughts without content are empty. The first half of the statement is directed at the radical empiricists from Bacon to Hume, and possibly at Newton's own "I make no hypotheses," and it says that their proscription of abstract ideas not only subverts discursive science but also destroys the unity and integrity of experience which they are defending. The second half of the aphorism is addressed to the rationalists, including Descartes and Kepler, and it says that hypotheses that are merely occasioned by facts and rush into systematic generality without any further empirical check let in the whirlwind of dogma, myth, and systematic illusion. Kant is here referring back to an ancient teaching that is first alluded to in Plato's allegory of the cave and in his account of the proportions of the divided line, both of which indicate that the hypotheses and the facts of experience faithfully reflect each other. Something similar is implied in the scholastic dictum that there is nothing in the intellect that is not first in the senses. Kant is issuing the same critical warnings in a new context where mathematics as well as imagination must enter a mutually disciplinary collaboration. Kant is saying that for the sake of both sanity and science the findings of experience and the intelligibilities of ideas must be constantly and continually responsible to each other. This is a tough mind laying down a hard discipline for both parties to the controversy. It is also a little puzzling that it comes from one of the coauthors of the nebular hypothesis, one of the more spectacular speculative inferences made from the law of gravitation, but Kant and Laplace are neither the first nor the last to break rules of method that have been laid down by themselves.

There are really three parts of this charter for the critical pursuit of knowledge. Each revises and re-establishes the role played by a part of science. The stuff of science is provided by the faculty of sensibility, as Kant calls it in traditional terminology. In the ancient world including the scholastic period, sensation was the process by which the mind receives impressions from bodies. With many variations and degrees of subtlety the process of sensation was compared to the impression that a seal makes in wax, the sensed body being the seal and the sense organ being the wax. There seems to have been little doubt that the body was a real object in nature, and that it acted as cause of the impression.

In the more subtle accounts, the resulting impression varies with the various properties of the body and with the various sense organs affected. In all these theories the impressions were identical or similar to their causes in the bodies. The trustworthiness of these data was expressed in the scholastic formula that there is no error in sensation. On the other hand there was always a doubt concerning the adequacy of sensation as a kind of knowledge; sensations were highly selective and ephemeral, partial and fleeting. Sensations, like all other things that change in time, are always coming to be and passing away; they never are. This doubt indicates that sensations are only parts of knowledge, necessary but not sufficient, always depending for their status on their interpretation in a larger whole. Their final validity depended upon their functions in relation to hypotheses and principles.

The attack of the empiricists on what seemed to them to be an unjustified dominance of speculative and analytic dogmatism cut sensations off from both their causal origins and their rational interpretations. Kant agreed with the skeptical criticism of the dogmatism, but he found Hume's fragmentation of human knowledge intolerable. His own reconstruction was made, however, within the limits set by the empiricists. The sensations are phenomenal merely, having no known causal origins nor valid objects outside human experience; but they are ordered by intuitions of space and time, and are therefore intelligible as themselves constituting the bodies which mathematics and physics describe. Nature is not a thing-in-itself nor even a system of things-in-themselves. It is the mathematical organization of human experience; human experience has become the subject matter of mathematical analysis; inversely, it is the system of systems.

The criticism of the rationalist is equally drastic and systematic in its results. For Kant reasoning by clear and distinct ideas is the systematic invention of fiction, myth, and illusion, and the cause of their systematic deception lies in the nature of the ideas of pure reason themselves. Ideas, separated from sensations and left to themselves, like the clouds of the Aristophanic comedy by that name, or like the gases of latter-day kinetic theory, tend to unlimited expansion. They have an appetite for totality, and this appetite exhibits itself in the infinite series that they generate and then try to comprehend. Such is the series of points or intervals in space that overleap any boundaries imagined even by children when they search in their minds for the farthest star; likewise the series of causes that leads back to but cannot grasp the first cause; or the series of determined acts that tries but always fails to catch and smother human freedom. Aristotle had called universals "the wholes," and abstract ideas are now said to determine universes of discourse. Kant, following the kind of reasoning which the Renaissance had modelled on analytic mathematics, shows how all such reasoning with these hungry ideas of reason constructs imaginary but impossible worlds, ending in contradiction, antinomies, or paralogisms; in effect, he is saying that the possible worlds in God's mind are impossible because they make anything like God's mind a mad tangle of mutually destructive, because mutually contradictory, systems. At the same time, as in Leibniz's speculations, they tend to become things-in-themselves, and their deceptions therefore become double logical paradoxes and metaphysical monsters. Kant is using the systematic paradoxes on which dialecticians had ascended to heaven to show that systems themselves are impossible. Speculative ideas are inherently partial aspects of the objects in experience; when they separate themselves from their context in human experience and set up whole and independent worlds on their own, they are parading in disguises. The rule with respect to them is to avoid indulgence in such speculation.

On the other hand abstract ideas are potentially in the human mind and realize themselves whenever experience presents its stuff of intuition and sensation. Their proper use is to order and reorder such material. Their proper natures and roles are realized in the rules they provide through the schemes of space and time for the intelligible unification of experience. Abstract unity finds itself in the unity of a manifold which gives rise to finite quantities and degrees of intensity in quality measured against the dimensions of space and time in experience. The infinite series of causes conceived in the cosmological speculations of the theologian is fulfilled step by step in the search for temporal order in the items of experience. The ratios and equations that seem to report the existence of crystalline spheres or the routes of angelic intelligences are really only the graphs by which we plot the successive positions of the planets.

These two critical themes, the first showing that sensations constitute the material of mathematical forms, the second that clear and distinct ideas are essentially rules giving order to sensations, imply the third: that there are limits to the proper exercise of human knowledge, boundaries beyond which reason and therefore science do not safely go. These boundaries fluctuate with the accumulations and ingenuity of imaginative rearrangements of experience, as well as with the successes and failures to find the rules that expand the abstract ideas, but the general definitive formula for the proper use of reason is not to trust evidence without inference and not to trust inference without evidence.

Kant knows that this observational caution and intellectual asceticism are extreme, that the acquisition of data will at times outrun hypothesis and that hypothesis will at other times reach beyond the facts, and he not only condones such sins, but he encourages them insofar as they stimulate and guide the search. But he does condemn the uncritical acceptance of the things-in-themselves that these separate disciplines lead to when their mutual responsibilities are ignored.

It is difficult to give the full rational explanation of this charter from the hand of Kant so that it loses its air of dogmatic criticism and stultifying caution. It would be simpler if there were strong evidence in his writing that he was forgetting or ignoring the evidence that the human mind belongs in the realm of being, that the sensations are a form of real knowledge and that the intellect is "the faculty of being." But there is no such evidence; on the contrary, there is much evidence that he is fully aware of the speculative powers of reason by which principles and hypotheses had been discovered, and without which Newtonian science, which he is defending, would not have existed. There is evidence that he was quite happily capable not only of following the physical speculation of his time but also, as in the case of the nebular hypothesis, of outdoing its boldest achievements.

But his strong theme is critical; he seems to be trying to save reason from speculative suicide, perhaps from the tragic *hybris* which he sensed in the brilliance of the Newtonian synthesis. He knows that reason is that power in a man whose reach always exceeds its grasp; reason inveterately asks questions of itself that it cannot answer. When it tries to answer such questions, it rushes headlong into pretentious dogmatism and eventually ends up in disillusionment and fear of itself. Kant saw this happening not only in the scholastic tradition of his time but also in the great hypotheses of his immediate scientific predecessors. In the book called *Prolegomena to Any Future Metaphysics,* he issues the warning: "All metaphysicians are therefore solemnly and legally suspended from their occupations till they have answered in a satisfactory manner the question, How are synthetic cognitions a priori possible?"[24] He himself answers this question for mathematics and physics by showing that synthetic cognitions a priori are achieved in judgments that combine concepts with intuitions of space and time so that one does not outrun the other, and these become the standard by which he shows that metaphysics and much science violate the rule. His own critical argument takes the form of showing the *reductio ad absurdum* of all other judgments. It is these negative critical conclusions that leave the reader of the *Critique of Pure Reason* suspended in a dream of physics that seems to consist exclusively of aesthetics and mathematics.

The reader who is acquainted with the mathematical physics of the twentieth century will not be surprised by this feeling of suspense in the medium of evanescent observations and oracular equations, but he may be surprised that Kant has anticipated so much from such a distance in the past. Kant has some of the prophetic quality that is often associated with Plato, and he has this quality for somewhat the same reasons. Plato's writings, viewed at full length, show pretty conclusively that his deepest and most persistent interest was in law, and that he viewed science—particularly mathematical science—in the context of law and politics. The same is true of Kant. Both men were impressed with the great scientific achievements of their times, and were disturbed

[24] Kant, *Prolegomena to Any Future Metaphysics,* trans. Paul Carus (1902; reprint ed., Chicago: Open Court Publishing Co., 1949), p. 29.

by the threatening admixtures of sophistry that accompany the scientific enterprise. They were driven to a search for wisdom, and found comfort at least in a method of science that imitates the methods of law. Each made a kind of myth of his conclusions. Plato thought there was no hope for human society unless philosophers became kings and kings philosophers; Kant looked forward to a time of *perpetual peace* when the operations of a world government will be responsible to the inherent wisdom of the common man. It should be added that both know the mythical character of their solutions.

Kant's prophetic power comes from his interest in the fate of the common man as well as from his acute analysis of what science was doing. He regarded the pure speculative pretensions of science as due to the irresponsible pride of the intellect. The solemn and legal suspension of the occupations of the metaphysicians and the scientists who unawares had turned metaphysicians, which he announced in the *Prolegomena,* sounds like the injunction of a judge in the republic of learning calling on the geniuses and heroes of the scientific conquest of nature to return to legally responsible procedures. The tribunal from which this injunction is issued is the court of common human experience, and the procedure imposed is to keep within the boundaries of its jurisdiction.

It is clear in the context of the *Critique of Practical Reason* and the *Critique of Judgment* that the procedure is not merely the submission of hypotheses to the evidence of the senses. Human experience is not merely the passive reception of sensations; it is a spontaneously active affair involving practical transactions between men and the conditions of their lives. It contains processes of doing and making in which reason constructs and leads a search; it is the exercise of practical reason. Imagination shows this, and without it there would be no rearrangement of the items of experience to accord with understanding. The so-called a priori forms of intuition are spontaneous, constructive, artificial. Men also exercise the productive arts, both in the handicrafts and in manufacturing. Kant was aware, as was his contemporary Adam Smith, of the use of mechanical tools in the early stages of the industrial revolution. He was anxious that the commercial arts should serve the common good as well as private greed; the arts of the legislative assembly and of the executive magistrates, and the new democratic functions of the republican citizen in England, France, and America, moved him very deeply. In all these activities in human experience he saw reason as the maker of rules, the lawgiver, the legislator. The deep root of this legislative power of the human mind is in the appetites, or the inclinations as Kant chose to call them. The formal authority of the power is in man's moral sense, or the rational will.

Corresponding to these powers and their activities there are habits and associations, within men as individuals and between men as members of communities and institutions, and the authority of practical reason issues imperatives that go with them. If one has a private end, he must have the skills that will achieve the end; if he needs to work with machines or in institutions, there are rules to follow; for all members of the human race there are general rules without which man's efforts are senseless and ineffective. The most general rules are given in three formulations of the so-called categorical imperative: 1) the rules of men's actions should be universal; 2) each man is an end in himself; 3) all of nature should be considered a kingdom of ends, or a system of means and ends. Although Kant never uses the poetic expression, it is impossible to read these rules with understanding without recalling the City of the World, and the Cosmopolis of the Roman Stoics, Marcus Aurelius and Epictetus. Man is a citizen of the universe and the categorical imperative is its constitution, the law with which all other laws must accord.

Viewed in this way, it appears that Kant has returned to the natural law of the Stoics and of the Roman Empire, and that he has identified the scientific laws of nature with political laws, or perhaps has substituted moral law for speculative systems, but this interpretation of Kant would let him fall under his own critical condemnation of uncontrolled speculation. Actually he is following later political theory, the theory of law to be found in Rousseau's *Social Contract,* in which natural law figures merely as a mythical background for the making of civil law by men. In the *Social Contract* men are imagined to have assembled to draw up an agreement, an agreement as to how they would make further agreements. Rousseau tells the story as if men agreed to give up their rights, or their power of making privileges or private laws for themselves as individuals, in order to make common laws in obedience to which they would find both their freedom and their rights secured. Kant's imperative seems to be a spelling out of the substance and implication of the agreement, and also an extension of them into the mind of the individual man. The individual attains his freedom when he makes good laws for himself, and the mark of a good law is its universality, since universality insures equality of consideration and regulation of all of his interests and inclinations. Likewise freedom of association and collective action is assured by general political laws. But the third formula of the categorical imperative pushes the Rousseauan doctrine of freedom farther than its author saw; man legislates for nature in the organization of his own experience. Nature is the result of man's ordering of experience in his understanding, and if he is to learn wisdom from science he must understand nature as a system of means and ends. To get the full view of this revolutionary interpretation of science, one should recall that the Stoic law of nature was the result of legislation by Zeus, and this logos doctrine had been translated into Christian terms where natural law is just subordinate to divine law and just above civil law which imitates it. Kant's natural law is made by man out of the materials of experience, sensation, and inclination, regulated and ordered by reason. Natural law, including scientific natural law, is made by man's rational will. The validity of science arises from the highest authority that men know, their own consciences.

In mid-twentieth century, this hundred and fifty year old doctrine seems loaded with almost too much fateful weight. Starting, as it seems to do in Kant's writings, from a philosophical concern about the validity of science, it involves the deepest principles of modern politics; it can be taken as the comment of a philosopher who watched the French and American Revolutions with liveliest interest and deepest concern. It finds the validity of science resting on a faith in the moral sense of men; but it also implies that the good life both of individuals and of humanity depends upon the integrity and success of the scientific enterprise. The historian of the last hundred and fifty years may not like to admit that the program of aesthetics, mathematics, and morals laid down in the critical philosophy is the adequate rationale of his story, but he would find it difficult to omit it from his account. It is not altogether a happy story, and Kant is not altogether happy in his exposition of its program.

Kant's unhappiness is shown partly in his style, which combines German grammar with the Latin words of the vanishing common language of scholars, which was slipping into the technical terminology of science. This produces an effect quite the opposite of Shakespeare's writing, in which the infusion of the learned language into English three centuries before resulted in the brilliance and clarity of great poetry. Kant's language is like a thunderstorm; it is heavy, dark, and threatening on first reading, and then it delivers thunder and lightning with consequent dramatic visions of a new and immense world. On repeated readings it has the sublimity of an epic which overpowers the imagination, forces the reader to high abstraction, and allows him as from a great height to look down upon a storm-wracked earth, "where ignorant armies clash by night." This is the meaning of transcendental criticism, Kant's name for his method.

But as usual the troubled prose style is symptomatic of deep trouble in thought, and Kant called his attempted cure the "Copernican revolution in thought." This is often taken to refer to the shift from the so-called objective realism of the scholastics and the rationalists to the subjective idealism of the empiricists, but that is a mistake in analogy.

Kant is really carrying through the logic of the analytic method, cutting it loose from its half-forgotten roots in metaphysics and theology and assimilating the new philosophy of law, which had also freed itself from theology. The categorical imperative is the result of the intuitive induction in the field of morals, and it becomes the sun around which all the planetary activities of men swing. But the old habits of thought disturb the activities of men, as large comets and meteors are said to have disturbed the early solar system. As Kant says, human reason runs necessarily to the speculative infinities and totalities. These innately rooted excesses of reason, exercised freely in the speculative manner, lead to conflicts, contradictions, and perversions of reason; at one stage they are called problems of reason: the existence of God, the immortality of the soul, freedom and determinism, the infinitude of the world or the universe. Reason formulates them, but it destroys itself in any thorough attempt to answer them.

It is interesting that Kant calls these problems which give rise to pairs of contradictory arguments antinomies, illegalisms. This suggests his solutions to the problems: turn the ideas of reason from which they arise into rules—the laws of scientific method—to reduce manifolds by measurement to unities and magnitudes, to trace causal sequences, to find the unifying threads of serial order in temporal experience, and finally to treat all experience as a realm of ends in the jurisdiction of reason. But Kant knows full well that this rule of reason is a process in time, and that its government is never stable since the antinomies work as a ferment, reason appearing repeatedly as the justifier of wars and revolutions for the sake of more reasonable peace. He knew that it did not simplify or smooth the road to remove heaven to its indefinite if not infinite end.

The conflicts of laws are to be found in governments and institutions, in the moral relations between individuals, and in the depths of the souls of free men. They are also found in factories, laboratories and observatories, as well as in the Associations for the Advancement of Science, all of these the institutional offspring from the mind of Francis Bacon, the Lord Chancellor. It is not exaggeration to say that all the problems that arise in the speculative use of pure reason are to be found in the modern complexes of science and law, no matter how closely these diverse expressions of the common logos are bound together, no matter how comprehensive their applications are made.

Perhaps it is the better part of wisdom to retreat to a metaphysical point of view before we review the effects of Kant's Copernican revolution in thought. It was pointed out earlier that the copula *is,* in the formal logic of Aristotle's organon, is not merely the sign of predication, the connection between the subject and predicate of a proposition;

it is also the sign of truth, the sign that the proposition purports to be reflecting the real being as it is. The fact that we judge some propositions to be false or merely hypothetical, therefore pretenders or imitations only, does not cancel the obligation; rather it re-enforces the necessity of considering and judging the truth which is in question in any discourse. This is the essential end and principle of the scientific enterprise.

The revolution instigated by the manifestos of Francis Bacon, joined in practice and comment by many scientists and philosophers in the following three hundred years and finally announced as accomplished by Kant, is based on an appeal to another end and principle of science, the good. It is the principle that Plato has Socrates invoke in the *Phaedo,* the principle of intelligibility which saves science from speculative futility, and again in the *Republic* where it is said to be the source of all reality and value. It seems to be a principle of metaphysics as well as of human affairs, where it appears in the *summum bonum* of human happiness, in the common good of the political community, and in the particular and apparent goods of the individual.

The mystics report that the true and the good are ultimately resolved in an unbroken unity, but philosophers, who for the most part fall short of the mystic vision, can report only partial identities in overlappings and intersections of the two principles which obstinately distinguish themselves from one another. The scientist can admit that the truth of his findings is a good of the intellect; the statesman will admit that the strategy of the truth is best in the long run; the ordinary man can see that science is useful, but he also is painfully aware in his day-to-day affairs that the true and the good are not simply the same, and his conscience confirms the difference between them. It is either a very wise man or a very foolish one that sees in the advancement of science an unmixed good.

Kant saw the antinomy in the application of his own solution to the science of biology. Ever since Descartes the revival of the ancient controversy concerning final causes had engaged European thought. Descartes had argued that animals were machines, and had brought much evidence from dissection and from physiological analysis to back his argument. It was fashionable to construct working models of animals to be run by water pressure from the fountains in the formal gardens of the seventeenth century. These mechanical animals were thought to be instructive as well as amusing illustrations of the arguments that were carried on in the witty society of the day. They were the ancestors of the Frankensteins and the homunculi of the eighteenth and nineteenth centuries. Kant, with the Newtonian persuasion of the validity of the science of mechanics, was sympathetic with the mechanical theories of biology, but he also discerned in the living organism the operations of a set of reciprocal causes that empirically conformed more closely to the concepts of means and ends than to the concepts of mechanics; in fact, he went so far as to repeat emphatically that not even a blade of grass could be conceived to have been produced by the action of mechanical causes alone. And yet the complicated concept of reciprocal means and ends within the organism yielded not an explanation, but only the analogical suggestion of an explanation, a suggestion that nature is in each of its products an artist making itself, a self-fabricator.

Kant recognizes in this the logical tangle, the infinite circular process, which marks the unchecked use of pure reason, an idea that runs itself to an infinite totality. He therefore applies his regular technique of reducing the idea to a rule: the discovery of an objective purpose in nature directs the scientist to search out mechanical causes in living organisms, and these can then be recognized as means to the objective ends that the purpose indicates. This process will be necessarily incomplete because of the limitations of the human mind. Biology will therefore be the method for discovering or making more and more laws of nature, but the biologist will always be left with a remnant of wonder and admiration with respect to the life of the organism which is radically beyond his powers of scientific understanding. This central and comprehensive biological problem is also central and crucial for the combination of science and law that Kant's method calls for, as well as for the dialectical resolution of the two metaphysical criteria, the true and the good, to which modern science appeals for its justification.

It may seem to the reader that Kant, even though he was a scientist as well as a philosopher, has been arbitrarily accepted as the proper judge of the issues of modern science. But the function of prophecy as well as criticism is sometimes thrust upon judges, and there are points of prophecy by which in time the quality of Kant's judgment can be tested. The remaining great books to be discussed in this essay, a good deal more scientific writing that cannot be included—some of it great and some of it not so great—and the general effects of the diffusion of scientific knowledge and technology of which no living man can be unaware, all these can be taken in evidence for or against the justice of Kant's criticism and prophecy.

The Kantian Program

At the start the three points of Kant's program can be accepted as the three great themes of the time, empirical investigation by observation in the field and experiment in the laboratory, political innovation and revolution in the law to parallel the industrial revolution, and incredible construction in mathematics applied to all aspects of life. The greatest feats of intellect with the most widespread effects occur in mathematics, but as would be expected from Kant's program, this inspires and forces the organized search for facts and evidence to support it. Both of these combined in technology pose the most grave and formidable problems ever faced in science, politics, and law.

Ironically enough, reports of some of the most fateful developments are unintelligible to the common reader. There is a reason for this. For instance, one of the tributary developments that in the last generation appears to have colored and dominated the main stream of thought is in chemistry, which, as its traditional name indicates, is the art of pouring, or more figuratively, the art of qualitative transformation. It is an applied or operational science which borrows its theoretical base from other sciences. It is prefigured in the transformation of the elements, first by Plato in the geometry and arithmetic of the four regular solids that represent the four elements—earth, air, water, and fire. Of these, fire is the most mobile and fluid, and it attacks, reduces, and reconstructs the other three according to definite formulae of combination. In Aristotle the regular solid atoms are dropped and the formulae describe qualitative change, a theme that gets thorough operational treatment in the sequel, but practically no theoretical development; the dark science of alchemy is the result. With the revival of Greek science in the Renaissance the beginnings of a theory appear in the designation of salt, sulphur, and mercury as elements, or solid, gaseous, and fluid states of matter, as we would call them, and in the use of fire as the transforming and explorative reagent. Chemistry became the art of the furnace for assaying materials. In Robert Boyle's *Skeptical Chymist* there is a report of an obscure and painful crisis in which the systematic use of the chemical balances demonstrates the essential differentiating property of weight for the elements. This is the true origin of the modern application of mathematics to chemistry.

Lavoisier's *Elementary Treatise on Chemistry* is the clear and eloquent account of his equally clear and systematic operation of the furnace and the chemical balances. Through these operational techniques a highly precise and consistent qualitative language is matched and meshed with the clear and distinct ideas of mathematics. Chemical analysis gets its mathematical as well as its operational meaning. The central traditional phenomenon of chemistry, combustion, is referred to the weighted and measured element, oxygen, instead of to the artificial fiction, phlogiston. From this start the progress of chemistry is assured in the application of the same techniques to the infinitude of qualitative differences and changes as they are noted in nature and produced artificially in the retort. The imitators of Lavoisier continue to differentiate the stages of the transformations by introducing the balances at every step until twenty elements are isolated and identified by numerical weights correlated with qualitative properties.

By this time enough records of transformation have accumulated to show that the compounds of these elements occur according to a regular pattern that can be expressed in constant numerical ratios. John Dalton then brings up the arsenal of mathematical possibilities in the analytic systems and makes some progress by their application toward a system of chemistry, the title of his book. At the same time, half in imaginative play with symbols and half in serious revival of the system of Democritean atoms, he invents the chemical atom, its varieties corresponding analogically with the atomic weights. This again accelerates the discovery and isolation of more elements and the application of more formulae for the compounds, until a cyclical order is noted in the series of elements that has resulted from the atomic weights. Mendeléyev discerns the cycles marked by recurring properties and expresses them as a matrix of rows and columns or alternatively as a spiral of ascending numbers. From this he predicts the discovery of new elements, and his predictions are still being fulfilled.

Faraday discovered the essential role that electricity plays in the art of pouring; lightning appears to be the fire that the flame of the Bunsen burner kindles in the dark cave of matter, and the chemical equations express darkly the electrical composition of bodies. Chemistry then begins to play an important pervasive part in all the sciences—physics, biology, and even astronomy—and is finally caught up in the theories of thermodynamics and electromagnetism as they are formulated in the high analytic arts of Willard Gibbs and Clerk Maxwell.

This intentionally brief account of the assimilation of chemistry to analytic mathematics is in part a simple illustration of the method that Kant formulates. Observation and experiment sort out the temporal sequences of qualities in the processes of combustion. In these complicated patterns, which show both sequences and combinations of qualities, some qualities persisting or recurring, others appearing and disappearing, Boyle's guess and Lavoisier's guiding insight that weight is the thread upon which all the other qualities can be strung is the key to a kind of perspective in which the terms of the formulae of combination and transformation not only record qualitative changes, but identify substances and assign properties to them. It is worth emphasizing here that Lavoisier throughout his investigations accepts the principle that the total weight of materials undergoing change in his experiments remains constant, and that he sets up controls in his experimental apparatus that conform to this principle. This principle seems merely to assure that the experiment fits the equation, but it also leads to the acceptance of the conservation of matter as a principle. So far the Kantian rule that mathematics fits with observation is obeyed.

But subsequent developments in Lavoisier's procedure connect with a major hypothesis that always and apparently inevitably goes beyond observation. Dalton, at first skeptically and then with increasing confidence, supposes that the chemical materials are composed of small elementary bodies. The equations composed from the constant ratios of combinations describe not only the observed constant weights, but also the unobserved motions and combinations of atoms, and this theme of equations of motion describing the relative motions of bodies persists throughout the nineteenth and twentieth centuries. Sometimes the bodies are gross, large enough for observation, sometimes they are macrocosmic, too large for observation, and sometimes they are submicroscopic, too small for observation. Very often these critical distinctions, so crucial for empirical method, are ignored partly because of the telescope and the microscope, and partly because the equations of motion sweep bodies of all kinds into a spectacular system, what Bacon would have called an Idol of the Theatre. In Kant's terms reason perversely forces the imagination into a world system, or even into a system of possible worlds. Today we would say that the mind tends to become a cosmic totalitarian dictator. At any rate the mind strains at the leashes that Kant tried to put upon it.

In the great books that report these world-conquering adventures, the dominating style is analytic mathematics. Many of them have been called mathematical poems; all of them might be called symphonic developments of Newtonian themes. Some of them are elegant and sustained developments of the central theme of gravitation in the *Principia,* and some of them are imitations of that theme in other mediums.

Many of these books are contemporary with Kant, and they could not have failed to impress his mind, or even seduce it, as seems to be the case in his involvement in the construction of the nebular hypothesis along with Laplace. Laplace's great book is the *Celestial Mechanics,* a complete systematic mathematical treatise which makes an exhaustive application of the gravitational law to the solar system. Another contemporary of Laplace, Lagrange, who invented a new branch of analytical mathematics, wrote *Analytical Mechanics.* Fourier transferred the analytic technique to the empirical findings in the field of heat and by means of the mathematical series that go by his name was able to reduce all the phenomena of conduction, convection, and radiation of heat to differential equations. *The Analytical Theory of Heat* shows the world as a system of heat. Clerk Maxwell, using the empirical findings of Faraday in electricity and the work of Young in optics, constructed a few differential equations that mirror the world as a system of electricity and magnetism. All of these are supreme products of the analytic art of mathematics, done by artists endowed with intuitive and deductive powers that would have amazed Descartes, and sustained by an economy and comprehensiveness of symbolism that rivals if not surpasses that of the great poets, Homer, Dante, and Shakespeare. The systems are complete and closed so that they seem to be successive concentric spherical networks that nest and reflect the empirically known world like the crystalline spheres of the ancients. Gravitation, heat, electricity, and light are the semitransparent luminaries that silently revolve between us and the fixed stars.

Even though the man in the street is only vaguely aware that these hypothetical systems exist, he is inevitably aware of many of their quasi-astrological effects. As the passage of a star over the meridian sets our mechanical clocks and sends us on our daily duties, so these mathematical systems demand the search for the special kinds of sense data that will support their rule; hence the importance of the laboratory and the field research expedition that have made so much news and so much history in the last hundred years. The equations demand and direct the research; then they sort the data, separating the illusions from the facts and distributing the established factual product to its proper category and system. The anomalous fact that persists starts the mathematician off on another construction that will divide it and continue to rule. There is something like the race between the hare and the tortoise in the relation between the mathematician and the experimenter, but it does not yet appear which is which. At any rate there is a Kantian agreement, or truce, between them that they are necessary to each other and must eventually agree in the republic of learning. But there is also a kind of conspiracy between them against the critical philosopher. They agree to a collaboration in the production and support of superfluous hypotheses, hypotheses that violate the rule that goes by the name of Occam's razor, to make no hypotheses that are not necessary.

These hypotheses for the most part are bodies and ghosts of bodies, which act as obedient servants of the equations, and as the substance of things unseen for the empiricist. The celestial bodies give out light signals; the gross or molar terrestrial bodies exert resistance to manipulation, and these two kinds of bodies together justify Newton, Laplace, and Lagrange in setting up the laws of motion. Molecules correspond to the older corpuscles, and they can be seen in their effects in Brownian movements, as well as supposed in the gases, liquids, and solids of the theory of heat, stated analytically by Fourier, and more precisely by Willard Gibbs. Atoms, as the component parts of molecules, answer to Clerk Maxwell's laws of electricity and magnetism and make a physical theory for observed chemical affinities. Later, laws formulated for radiation give birth to the subatomic particles. Each sphere of theory has its own elementary bodies, and all spheres working together populate a world which is known to the man in the street as the material and mechanical universe. It is not surprising that the mid-nineteenth century saw what it thought were the closing episodes of a world conquest by the scientist; there were only a few confused corners to be mopped up.

But Kant was right. These material bodies, these things-in-themselves, whether bodies or equations, were merely temporary achievements in the organization of human experience. Even as the equations of motion identified themselves with the eternal laws of God in the theology of Deism, or with the laws of the independently existing nature, both the empiricists and the mathematicians were busy finding new anomalies and new laws to meet them. The phenomena of radiation were symptoms that the elementary bodies were imaginative children of mathematical hypotheses that could be replaced on demand. New equations were made for new observations, and in the depth of mathematics new foundations were being built. Lobachevski and Riemann were constructing new geometries alternative to Euclid's, in which there were no parallel lines or many kinds of parallel lines that Euclid never imagined. At the same time Dedekind and Cantor were finding new numbers to meet the growing demands of the calculus and reconstructing the whole number system to assimilate them. It was even necessary to look for a new logic to take care of the principles that govern the application of mathematics to new and apparently inconsistent data.

It is interesting to note that all this new construction in mathematics is made on the basis of postulates rather than axioms and common notions. The premises of mathematics are taken by a process of rational demand, rather than by self-certification. It seems that the new queen of the sciences acts as a constitutional monarch preferring tentative injunction to edict. No mathematician seems to have paid much attention to Kant, but this method of postulation may in spite of that be a symbolic surrender to Kant's canons of legislative validity. The heaven of mathematical forms is also ruled by the categorical imperative.

It is a cause for shame, and almost a public scandal, that the great books that report all these doings in the republic of learning cannot be read by the common man. The basic reason for this is that our public teaching of mathematics, the discipline more than any other one that rules our lives, is of such low power and poor insight, that the ordinary schoolboy is shut out from the mathematical part of his world for the rest of his life. It is possible to study mathematics so that the mathematical poems, and even some less elegant treatises can be read as literature with no more pain or difficulty than other classics of Western thought. In spite of much educational talk to the contrary it is now true, as in Plato's day, that the man who ignores or is ignorant of mathematics cannot enter fully into the life of his world.

Faraday

Two men of science who were comparatively innocent of mathematics but whose influence nevertheless entered very deeply into the life of the world of their and our times did write great books that can be read. They did not deliberately secede from the jurisdiction of mathematics; they simply found themselves involved in experimental and observational investigation for which their clear and forceful use of ordinary language was adequate.[25] Like Gilbert and Harvey, whom they respectively inherit and parallel, Faraday and Darwin reflect in their mode of thought heavy influences from the analytic arts which they do not themselves practice. Both accept simply and uncritically the materialistic world which had been precipitated from the great systematic hypotheses of the early nineteenth century. If they had been asked about its validity they would simply have pointed to the "evidence of the senses." They are empiricists.

Newton, at the end of the General Scholium of the *System of the World,* Book III of the *Principia,* says:

[25] A close reading of Faraday's *Diary* and *Experimental Researches in Electricity* suggests that Faraday does in fact secede from the jurisdiction of mathematics self-consciously and intentionally, aiming to construct not the groundwork of a mathematical theory, but rather a nonquantitative experimental science, a science of facts complete in itself.

And now we might add something concerning a certain most subtle spirit which pervades and lies hid in all gross bodies; by the force and action of which spirit the particles of bodies attract one another at near distances, and cohere, if contiguous; and electric bodies operate to greater distances, as well repelling as attracting the neighboring corpuscles; and light is emitted, reflected, refracted, inflected, and heats bodies; and all sensation is excited, and the members of animal bodies move at the command of the will, namely by the vibration of this spirit, mutually propagated along the solid filaments of the nerves, from the outward organs of sense to the brain, and from the brain into the muscles. But these are things which cannot be explained in few words, nor are we furnished with that sufficiency of experiments which is required to an accurate determination and demonstration of the laws by which the electric and elastic fluid operates.

This incredibly informed and comprehensive statement by Newton in 1713 suggested, if it did not actually direct, a program of experiments that were carried out in the next hundred and fifty years by many different men, some of whose names are honored in the technical terminology of electrical theory, Galvani, Volta, Ampere, Ohm, Henry, and Faraday. Faraday, in his *Experimental Researches into Electricity,* very nearly equals Newton in his power to grasp, supplement, and integrate a large collection of fragmentary findings by other men, and to thread his way through the details to the single general view. The great difference between Faraday and Newton lies in the almost complete absence of mathematics in Faraday's method, a deficiency which was filled later by Clerk Maxwell, who provided the literal elegance of the analytic symbolism. The harmony of Faraday's verbal and technical imaginations is almost a miracle of incisive clarity, particularly when it is compared with the prevailing mathematical styles of the period. It tends to dim the brilliance of Clerk Maxwell's work, which appears, quite falsely it must be admitted, to be the work of a mathematical clerk. By his deficiency Faraday loses in economy of expression and in unity of theory in the end, but he makes up for it in the completeness and thoroughness of his empirical labors.

There are few scientific writers who convey to the reader the feel of manual operation in the laboratory as Faraday does, and the profit in reading him is enhanced if the attention of the reader keeps Faraday's observations closely related to what he does with his apparatus. Everybody knows in general that experimental research has a large component of manipulation in it, but there is very little literature which allows the general reader to realize it. Faraday got his skill by working as a kind of apprentice to Sir Humphry Davy, who was a chemist, and his early work consisted in carrying out in operations the many ideas that Davy entertained, one of the most characteristic of which is embodied in the miner's safety lamp, the so-called Davy lamp. The principle of the lamp is the diffusion of heat through the mesh of a wire screen placed around the flame; the screen so spreads the heat that no single point of the screen becomes the point of ignition for surrounding gases.

Davy had a roving eye for the differences in materials and their properties and a quick mind to grasp those relations between the sizes, shapes, and qualities that fit useful operations. It has become commonplace to attribute this utilitarian aptitude of mind to the British and American inventor, and to credit the industrial revolution to such mechanical genius. Davy and through him Faraday are prolific inventors, but at their best they are unconscious artists, tending to see nature as the original inventor whom they merely imitate. There seems to be some deep affinity between the kind of imagination that arises from skill of hand and eye and the operations of nature that go with the theory of the electrical constitution of matter. All of this seems beautifully symbolized in the electrical engineer's hand when he holds it out before him to represent the polar axis of the magnetic field, the direction of the current, and the sense of rotation, the left hand with thumb and index finger extended at right angles to one another and the fingers clenched for the motor, and the right hand similarly disposed for the dynamo. These are two electric machines for which Faraday supplied the theory before they were designed by engineers.

The Experimental Researches in Electricity is full of such imaginative pictures which give quasi-bodily form to the "subtle spirit" of Newton's comment. Up to Faraday's time there were two kinds of electricity: frictional or static, made by rubbing such materials as wool and amber together, and evidenced in attractions and repulsions; and voltaic, made by two metals immersed in salt or acid solutions. Faraday demonstrated the identity of these two kinds of electricity by a very skillful substitution of the static machine for a battery, and made the crucial inference from the same apparatus that identified positive and negative electricity as equal and opposite charges on bodies within a single field. He also discovered electrolysis to be the process opposite to the generation of voltaic currents by the chemical dissolution of metals in acids, and inferred from this process the measure of the quantity of electricity in any closed system, the electrolytic tank thus anticipating the present familiar wattmeter. There were many byproducts of these necessary fundamental integrations of previous knowledge and know-how, such for instance as the catalytic powers of red-hot platinum in the combination of gases, or the motion of ions in the flow of electricity in fluids.

But the great discoveries of Faraday deal with the inductive phenomena, and in these the guiding threads are half-imaginative, half-abstract geometrical conceptions that are most characteristic of the constructive empirical mind. With the help of iron filings he made studies of the forces acting in the region of the fixed magnet. The positions and orientations of the filings suggested to him, as they do to any child, lines of force connecting the poles of the magnet in a three dimensional pattern of subtle curves. The phenomena of attraction and repulsion connected with these lines suggest elastic bands, and taken together the elastic bands make a kind of aura, or ghostly body, surrounding the solid magnet. The suggestion also is present that the iron filings are imitating the particles out of which the solid magnet is composed and that these particles align themselves along lines of force that pass through the magnet. All this, of course, had been prefigured in the behavior of Gilbert's terrelas, but the style is changed; the lines of force have taken the place of the loves of the lodestone.

Three different men in the year 1820 had noted the effects of electric currents on magnets. Oersted saw that magnetic needles take a position at right angles to the flow of the current in a wire; Arago noted that pieces of steel become magnetized when placed near wires carrying electric current; and Davy himself saw that soft iron gains magnetic powers when a current is passed through a wire in its immediate vicinity, and that such powers disappeared when the current stopped flowing. A helical solenoid, made by winding insulated wire about a hollow cylinder with ends attached to a battery, produces a magnetic field similar to that of a bar magnet, and this solenoid can be used to magnetize iron by induction. The term *induction* had been used before to designate the process by which the lodestone magnetized iron. It appeared from these phenomena of so-called electromagnetism that an electric current itself supported a magnetic aura or field, and the analogy was so close that Faraday seized upon the question of what the inductive relations between two currents might be. In a period of ten days he had made the relevant experiments and found the classical answers.

If two wires are parallel to each other, the change of current in one will induce a current in the other. In the same wire there is self-induction upon the breaking and making of the circuit, or in any change in intensity of the current. It is clear that the magnetic lines of force around a current become the medium for the induction of electric currents. It is also apparent that induction takes place only by a change of electric state, or what might be called electrical acceleration. It is also clear from subtle manipulation of delicate apparatus that there are tendencies to motion and actual motions of bodies, magnets, wires, and electrodes universally implicated with the electric and magnetic tensions, and that these three, electric currents, magnetic attractions and repulsions, and local motions of bodies are uniformly related to each other in the images of the two hands of the electrical engineer.

There is a philosophic tradition of admiration for the human hand. As far as we know, the original literary expression of it is in Galen's chapter on the human hand in his treatise *On the Use of Parts,* which analyzes the operations of the human hand in terms of its bones, muscles, nerves, flesh, and nails, to demonstrate the elegance and efficiencies of its highly functional operations. The hand is not only a well-built machine, but also a most superior tool in which nothing is wasted and every discernible feature serves a purpose. It is the tool of tools, giving all other tools their forms and purposes, a kind of second mind that searches out intelligibility in nature, and confers intelligence wherever it seems to be lacking. This chapter is often quoted as the classical argument for a teleological view of nature, sometimes by persons who do not know the sophisticated mechanical theories which form its original context. Paley wrote a whole book, *On the Hand,* which is an expanded imitation of Galen's chapter. Karl Marx quotes it in his account of the transition from manu- to machino-facture in the industrial revolution. Most of these imitations emphasize the part the hand plays in artificial making and the evidence which such operations provide for the intelligence of men and God, but there is in the original Galenic account an opposite and equal emphasis on the inverse relation and function, the hand as a medium for the acquisition and coordination of natural intelligence—in short, as a scientific instrument. A modern account, perhaps inspired by a Leonardo-like vision of scientific method, would draw the main features of a theory of manual induction from Faraday's notebooks.

Faraday himself is not much given to self-description nor to self-admiration; his eye is on the material and the traces in it that his operations leave, but the medium through which he sees and knows is the work of his hands in fashioning and using the instruments that he makes with such efficiency and grace. This is acutely obvious to one who tries to repeat his experimental researches, and fails time and time again with merely ordinary fumbling hands; the facts are simply not there for want of the hands to receive them.

One gets the full realization of the part the Faradaic hands play when he looks with the help of Clerk Maxwell over Faraday's shoulder. Faraday's gently flying hands leave the images of their motion in the air for the mathematical eye of Maxwell to catch, and as Maxwell catches them, they turn into mathematical images. It is true that many of Maxwell's mathematical images were constructed with his eye on other models, but he is quick to identify them in Faraday's operations, and to note them in differential equations. Maxwell was also a skillful experimentalist, and it is not frivolous to say that their minds met in the motions of their hands where the better part of their minds lived and worked.

But they had other common mediums, mediums in which their imaginations moved and which moved their imaginations until they were persuaded that what their hands did, what their imaginations saw, and the operations that their equations symbolized—all these superposed and finally identified with one another—were what nature did electrically. The focus of the fascination that worked this magic was the lines of force, seen first in the pattern of iron filings around the magnet, then seen in the cyclical lines around a live wire, and then in the various combinations of magnets and wires that produced the phenomena of induction. Faraday realized that the filings merely indicated tensions, which he likened to rubber bands.

The ensuing struggle between his imagination and his scientific conscience resulted in paper after paper "On the Physical Reality of the Lines of Force"; and these papers were in turn the beginning of a controversy that continued through the nineteenth century on the principle of action at a distance. Seen at our distance Faraday was worried about the "physical reality" of form, in this case the form of a motion, commonly known as the path of a motion. The problem is similar to the question of the reality of the life of a man when he is dead. It is often referred to the theologian or the metaphysician, but these worthies might well begin their inquiries by referring it back to the historian: what kind of reality has the life of Socrates or Napoleon? A man lived, acted, and died, and we today not only read about the "life," but we feel its effects whenever we try to think big thoughts or attempt big deeds. Our consciences tell us not to see ghosts. So Faraday saw the alignments of bodies that indicated motions, motions past and motions in the future, motions that had been and would be beautiful and orderly. He remembered what Kepler and Newton had done with similar paths of bodies. Unlike Newton he wanted to make a hypothesis, a hypothesis that would explain, without magic and superstition, why a child skipping a wire rope facing the north star generated a current in the rope.

His papers are inconclusive, but very suggestive to a mathematician who knows that second derivatives are very good indicators of the evanescent states of bodies at stations in their paths of motion. Maxwell studied the whole literature of electricity, and he also wrote papers on the lines of force. He reduced the motions to units and dimensions, and consulting his imagination with these quantities in mind, he saw tubes of force revolving frictionlessly against each other. The inductive effects arose when they cut or disturbed each other. Further, he conceived and imagined a field of force ordered and structured by these interrelations. These were the mathematical images that touched Faraday's incipient hypothesis of physically real lines of force and saved them from gross materialization by lifting them to a unit of space filled with forms of electrical tension or incipient motions expressed by differential equations of the second degree. The hands and fingers of the electrical engineer, positioned for the rules that indicate the direction of current, magnetic field, and force, will serve as suggestions to the uninitiated of the electromagnetic patterns that Maxwell saw in his mind.

Although it is Faraday's eye, hand, and imagination that impress his reader, there is something about his workmanship that suggests deeper traits of character. It is said that he belonged to a small Protestant sect in religion which was a branch of the very widespread pietism of the previous century. Kant is the best-known spokesman for the spirit of the cult, and Faraday would have subscribed to his skepticism with respect to high speculation in theology and his deep regard for the moral law. He was skeptical about the physical reality of the lines of force, but there are signs of a deep devotion to the moral law in his exhaustive and laborious tests of the many phenomena in which their effects might be found. It is due to these labors of Faraday that Maxwell's equations penetrate so deeply into what is now known as the constitution of matter. In the three states of matter, solid, liquid, and gaseous, and in the chemical distribution of these then known to chemistry, Faraday tried all things to find their varying degrees of tension with respect to the lines of electrical and magnetic force. The process of electrolysis which he discovered became his chief instrument of investigation, but the solids and gases which this process isolated incited him to still further experimental techniques. The result of this investigation was the recognition of the universal presence of the electromagnetic properties of all matter. Each chemical element and compound had its dielectric coefficient and its diamagnetic coefficient, and it was therefore and henceforth clear that all material bodies can enter and qualify fields of force, electromagnetic as well as gravitational. Maxwell's equations describe the constitution of matter and become the superformulae of chemical combination and transformation. Lightning replaces fire as the chief agent in chemical change.

But there were other contexts for this feat of world conquest. Newton's two chief interests, gravitation and light, and his theoretical accounts of them, had been models for Faraday's imitations. It was unlikely that they would fail to turn up in Faraday's workshop. As this is being written, it is announced that Einstein has written the equations that comprehend the relation of gravitation and electromagnetism. Faraday began the search for this relation more than a hundred years ago with his many papers comparing and contrasting the two forces; but he was Moses seeing the Promised Land from Moab. He was happier in his discovery of one of the intersections of electricity and light. In studying the diamagnetic properties of heavy glass, he discovered that polarized light is rotated ninety degrees in the focus of a strong magnetic field; the context for this discovery in fact and hypothesis allowed a daring inference, namely that light is an oscillating tension in an electromagnetic field, and therefore is subject to the laws of electricity. This should be counted as one of the great crucial experiments, for although it did not at the time close off the many false leads that have since been followed to dead ends in the realm of larger hypotheses, for instance the labyrinth of the luminiferous ether, it has remained, after investigation has removed the fictions, a steady signpost pointing to the new physical world in which fields of energy rule the waves of radiation.

It is difficult to give specific weight to the importance of Faraday's individual achievements. His relation to Maxwell is a mystery at the center of the question. In this brief account of techniques and materials it has been impossible to keep Faraday and Maxwell apart, as apparently they were separated in their actual work. They do not in the end eclipse each other, but on the contrary complement and enhance an interpenetrative light. Together they extended, shook, and redefined the world of Newtonian physics and are still doing so. Their influence is not merely in the realm of theory. The technology to which their physical discoveries give rise justifies Lewis Mumford in designating it as the third engulfing wave of the industrial revolution, electric power, chemistry, and metallurgy following and finally leaving behind the water power and wind power and the coal and steam of the two preceding waves.[26]

As far as method is concerned, several comments can be made that do not conform to the legends of the historians of science. Faraday is undoubtedly one of the great geniuses of modern experimental investigation; consequently he bent all his skill and intelligence to the accumulation of facts. The essential and quantitatively large part of his work was made possible by powerful and subtle manual dexterity. So much is generally agreed. But the actual operations of his hands were guided, perhaps partly subconsciously, but nonetheless effectively, by the large speculative hypotheses and principles of an earlier period and

[26] See Lewis Mumford, *Technics and Civilization* (New York: Harcourt, Brace and Company, 1934), Chapter V.

atmosphere of science. This is true not only regarding Newton's influence. The fitness of his findings to the symbols of the calculus and the continental world systems reflected in them argues strongly for the powerful inductive influence of abstract ideas on an imagination nourished and rendered receptive by active hands and eyes. One cannot argue that the movements of hand and eye were dictated or deduced from abstractions, but there is something dialectical in this methodical collection of facts in defiance of the speculation which actually defines the locus and informs the rules of selection and relevance. It seems that the shining firmament of the scientific renaissance had been translated into the moral maxims of a scientific conscience in Faraday's character. Kant's moral law and starry heavens are most intimately associated in Faraday's laboratory. The effect is a radical empiricism impregnated with hypothesis and principle by art and artifice.

Darwin

THE mark of the nineteenth century empiricist is the notebook that records observations. Faraday's notebooks are voluminous and highly instructive regarding laboratory procedures, but Charles Darwin did not stay in the laboratory. Instead he travelled. He did not bring the mountain of natural phenomena to himself; he let his instrument, a sailing ship, take him to the mountains, the sea, and the plains, and he kept notes on what he saw. He was an empiricist, but his genius consisted for the most part in letting nature take its course, even in letting the wind, as sailboats do, set the locus of his points of observation. He even allowed his sails to be filled with winds of doctrine, albeit he used a rudder, sheet, compass, and log to keep track of his position.

As consequence, Darwin's great book is filled with description of surfaces, and where the richness of difference and variation of the features of nature give out, his habit of attention to such a multiplicity of forms carries him beyond the observed to the infinitely small and the infinitely great and he travels backward and forward in time to extend the field for observation that mere space presents. In fact the constructive magic of time that he builds into his natural history hypnotizes the reader of *On the Origin of Species* so that he is temporarily unable to distinguish observation and hypothesis. The so-called evidences for evolution from anatomy, physiology, geology, palaeontology are all facts heavily processed by hypothetical temporal constructions and beg large questions which only an impossible historical verification could answer.

But this is perhaps the nearest to fulfillment that the empirical method has come. Principle and hypothesis seem to be ignored and suppressed only because they are so completely and exhaustively absorbed in the observation and the record. For it is true that there is no implicit abstraction that is not exemplified in observation, and there is no observation that does not find its determinate place in some hypothetical temporal serial order. Abstract arguments and criticisms about the theory of evolution are futile, because they are arbitrary, gratuitous, and irrelevant with respect to the author's intention, which is to arrange the facts of natural history so that they all can be seen at one time in one temporal perspective. Temporal patterns are very rich in variety and diversity, and in Darwin they are hungry for his filling.

The temporal patterns come from all sorts of sources and theoretical contexts: from geology, in which the ages of the rock strata are determined by fossils; from Malthus's study of population in which two time series, one for the generations of plants and one for the generations of men, show a difference which amounts to the progressive starvation of men; and from history of the breeding of domestic animals, which shows that time and human artifice are the parents of innovation. Geology explodes the familiar time scheme of the Biblical story of creation and gives millions rather than thousands of years for the development of life on the earth. In this larger interval it is more than possible that the distribution of fossils in the strata represents whole systems of life, each of which failed in the struggle for existence which Malthus sees as the inevitable result of the ratios of plant and animal reproduction. The present epoch of plant and animal life consists of a system built out of the remnants of past systems, survivals which show the action of some principle of selection analogous to the purposes and judgments of fitness that men put into their practices of animal breeding.

It was the filling out of this analogy that Darwin accepted as the assignment in his book *On the Origin of Species.* It is a variation on the very old analogy between art and nature so often expressed in the proverbial formula, art imitates nature, and so often inverted, nature imitates art, and accepted as the guiding rule for the scientist's investigation. Darwin is puzzled by the plethora of fact he has collected; he makes a study of animal husbandry; and then wonders whether nature is not a bigger and better breeder of plants and animals. The guess might have led him back to the theological analogy which would make God the superartist, as this sort of wonder at sublime magnitudes often does, but Darwin, whether he knew it or not, was a follower of another line of high speculation.

Actually Darwin expresses himself in the well-recognized phrases of the nineteenth-century agnostics. He does not know about God, His mind and His purposes, and as a scientist he has no need of God in his hypotheses. But in spite of this skeptical temper, he retains what has recently been called a natural piety, a respect for nature's magnitudes and elegances which impose the human obligation to explain and understand.[27] Again this might have thrown him back on the sciences of physics and mechanics, or even on the methods of statistical analysis which his followers in the science of genetics have employed. But he does not even start in these directions.

The deep root of his natural piety is in his respect for facts. The quality of mind is seen in this quotation from one of his letters: "I have steadily endeavoured to keep my mind free so as to give up any hypothesis, however much beloved (and I cannot resist forming one on every subject), as soon as facts are shown to be opposed to it." Though the whole quotation has become an article of the constitutional law of the experimental laboratory, the special Darwinian flavor of it comes from the clause in parentheses. Darwin's mind apparently had a power of spontaneous generation of hypotheses which became an irrepressible habit as the power was exercised in the context of an equally irrepressible appetite for facts. If one looks for the manifestation of this power in *On the Origin of Species* or for that matter in any other of Darwin's books, it appears not so much in the construction of time hypotheses, which make the hypothesis of evolution, and which are a bit labored, but rather in the only half-formulated hypothesis of adaptation. This subhypothesis is present chiefly in the terms of the descriptive language by which he reports his observations and by implication in the selection of the facts that he chooses as evidence. It appears that Darwin had an eye specially tuned to a certain relation of facts, and it is this relation of adaptation multiplied in many instances that becomes the typical ubiquitous feature of the evolutionary structure.[28]

[27] An account of the stages in Darwin's thought on religion—his shift first from Christian orthodoxy to a vague theism, and then from theism to agnosticism—is given by Maurice Mandelbaum in "Darwin's Religious Views," *Journal of the History of Ideas,* XIX (1958), 363–78.

[28] Some danger of misinterpretation arises from the separation that Buchanan here makes between "time hypotheses" and "the hypothesis of adaptation," for in Darwin's theoretical work it is often the relation between these two aspects that provides the key insight. The adaptations in which Darwin particularly delights are those that can be accounted for plausibly in terms of long-enduring temporal processes, where the available alternative hypotheses require unexplained coincidences. In his coral reef theory, for instance, he accounts for the

The immediate background in the discussion of evolutionary theory, which also reaches back into the eighteenth century, had made the idea of adaptation very familiar. It had two sources, one a semitheological myth traditionally going back to Galen, and the other a theme in metaphysics first fully formulated in Plotinus. The Galenic myth, probably borrowed from the Stoics, tells animal stories in which all living creatures like man share in the providence and purposes of an original maker or fabricator of the universe. Their growth and their movements are immediate evidence of goods desired, plans entertained, and means used to accomplish ends. Galen is sometimes very free and generous in his grants of intelligence and will to animals, and then turns critical like Kant and says that the nonrational animals of course only act as if the story were true. The eighteenth-century French biologists accept the minimal critical interpretation and use a terminology that keeps some of the myth. They are concerned to arrange the species and varieties of animals and plants to show an order of descent by the inheritance of acquired characteristics, and this involves some explanation of the appearance of new organs and the disappearance or modification of old organs. The animal or the plant makes these changes by responding to tensions or needs that arise inside the body or in relation to the environment. Lamarck, following similar speculations by Buffon, explained the whole organization of animals and formation of different organs by the four following laws:

1) Life by its proper forces tends to increase the volume of every body possessing it, and to enlarge its parts, up to a limit which it brings about.

2) The production of a new organ in an animal body results from the supervention of a new want continuing to make itself felt, and a new movement which this want gives birth to and encourages.

---

shape of atolls by assuming that there has been a general subsidence of the ocean floor and oceanic islands in certain regions of the Pacific Ocean; the corals, which can live only in relatively shallow waters, build upward and so avoid death. The alternative hypothesis in Darwin's time was Lyell's, which assumed that the corals built on the rims of submerged volcanoes. Darwin's hypothesis (confirmed in this century by drillings which show the material of the atoll at great depths to be coral rather than volcanic rock) has the advantage of not assuming a large number of submerged volcanoes with rims at just the proper depth for coral formation. Many illustrations could be given to show that temporal sequence and particular directions of temporal sequence play a crucial role in Darwin's account of adaptation; it is just here, and in his departure from the traditional archetypal view of organisms, that Darwin is most innovative. A defense of the radically innovative character of Darwin's work in these respects will be found in Michael Ghiselin, *The Triumph of the Darwinian Method* (Berkeley, Calif.: The University of California Press, 1969).

3) The development of new organs and their forces of action are in constant ratio to the employment of these organs.

4) All which has been acquired, laid down, or changed in the organization of individuals in the course of their life is conserved by generation and transmitted to the new individuals which proceed from those which have undergone those changes.[29]

One can feel the plausibility in this language, and can see the vivid moving picture of the hungry zebra in an African drought reaching for food on the trees and then passing on its stretched neck to little zebra-giraffes, but the quasi-observed facts leave that critical tension of mind that myths engender. This gives one meaning to adaptation; one might call it the active or dynamic meaning.

The Plotinic metaphysical theme gives a quite different explanation of the quasi-observed facts. It will be remembered that the One from superabundance of being overflows or emanates all possible forms, but matter is capable of receiving these forms only in a temporal order, not all together but one after another. But this means that in time all possible forms will be received and exist, and this coming into being will take place on various levels. On the level of life this will appear as the growth and variation of varieties and species, and the temporal order of the emanations will give the appearance of evolution. In fact it is from this traditional theme that the terminology of evolution arises in the eighteenth century. The evolution of species is due to the emanative pressure of the possible forms of being. The phenomena of adaptation appear then as the detailed view of the compossibility of forms, the possibility of the coexistence of animals, plants, and inanimate nature at any given time. This puts a determinism back of the apparent desires or loves of animals and sets the problem of evolution in a larger context where the task is to construct a system of nature according to certain principles and laws:

1) All forms are to be placed in a temporal order.

2) Those that can exist together are compatible or compossible.

3) Those that cannot exist together may exist at different times, one before the other, or at different places.

[29] Lamarck first enunciated his laws in the form here given in the second edition of his *Histoire naturelle des Animaux sans Vertèbres* (Paris, 1815), p. 151. The third and fourth of the laws had previously been given in his *Philosophie Zoologique* (Paris, 1809), I, 235.

4) The number of differences in all possible forms is infinite so that between any two forms there will be an intermediate form.

5) The relations between forms are either positively adaptive (useful) or negatively adaptive (destructive); there will result either survival or a struggle for existence between forms.

A man of Darwin's time, the mid-nineteenth century, could have worked on the problem of evolution with either of these sets of postulates, the Galenic or the Plotinic. Actually the mid-nineteenth century did not encourage this degree of explicitness in its assumptions. It conceived itself to be radically empirical, and the parenthetical clause in the quotation from Darwin is a slip of the tongue. Furthermore, although the Plotinic postulates are highly speculative, they have a closer affinity with the empirical frame of mind than have the Galenic. Numbers 4 and 5, the so-called principle of continuity, were and are still being assumed in the greater part of mathematics and physics, whereas the principle of teleology in the Galenic postulates was being opposed vigorously because it seemed to be on the wrong side in the war between science and religion.

Darwin's natural piety fits with the Plotinic postulates: facts are the appearance of forms in observation; infinite variations justify empirical extrapolation and the search for missing links; and adaptation and the implied fitness to survive explain the system of coexistent forms at any given time. It perhaps should be recalled that Lucretius had applied a similar formula to the whole existing world; it fits a scheme of survival in a world of pure chance. Many modern geneticists interpret Darwin's theory of evolution this way and proceed to apply their own mathematics of chance, the theory of probability. But Darwin's spontaneous hypotheses concerning adaptation, borrowing as little as possible from the Galenic-Lamarckian postulates and allowing him to keep close to the observed facts, take the place of chance and statistical mathematics. His most "beloved" hypotheses concern the wonders of natural adaptation.

The reader of *On the Origin of Species* is always made happy by running across the original statements of the hypothesis of protective coloration in plants and animals, the theory of sexual selection, and the correlation of variations with migration. These and many other stories in his natural history have become common knowledge and almost proverbial folklore, but there is in Darwin's recording style the steady, persistent operation of the adaptive insight. His genius seems to lie in the intuitive induction of elementary teleology. He no sooner sees an animal or plant trait than he sees a condition in the environment that is favorable or unfavorable to the persistence or elimination of that trait. There is no imputation of the teleological myth of conscious or deliberate purpose, but rather an almost sympathetic insight into the relation which has since been called survival value.

The relation itself is complicated and many-termed, as all instrumental relations are, and it is this ramifying tendency in the idea of adaptation that Darwin has to guard against lest it run beyond the observational base of the intuition. A trait or an item of the environment may be useful in one respect and highly disadvantageous in another; two traits together may give a resultant advantage that neither had alone. The only safety in generalization would be through a complicated systematic logic. Darwin has no such abstract system, as many critics of the theory of evolution have pointed out, but he has what many great scientists have had in place of logic, an adaptation of instinct to insight, a sure touch, as we see and wonder at it in men like Faraday and Aristotle. In spite of nearly a century of raking criticism and highly charged controversy, Darwin's understanding of animals and plants has remained unassailable. The geneticists who have gone beyond his theories to inheritance and environmental influences are still trying to catch up with the basic structure of the economy of plant and animal life, the structure that is built out of Darwin's elementary adaptations. It is this that gives the stable permanence that one feels in the sweeping hypothesis of evolution.

The battle between science and religion that has raged around the theory of evolution arises from some of the applied implications of the theory, particularly as they bear on the origin and descent of the human species. It has appeared in this battle that Darwin is to be counted among the critics and refuters of Aristotle. The reader of Aristotle as well as of Darwin will be able to make some corrective distinctions. Darwin, or rather Huxley, was certainly able to refute the opaque dogmatism of the defenders of the religious theories of the origin of man, and these dogmas descend by dark and devious routes from theological doctrine which also passed on some devalued Aristotle. But the doctrine of fixed species in terms of which the battle was fought is a little more subtle and complicated than the controversy would show.

The original Aristotelian distinction between genera and species was a part of the formal science of logic, an analysis of the relation of universals. Its application to nature is a part of material logic in the sciences of nature. Formal logic is largely a science of principles, such as the law of contradiction; the sciences of nature involve hypotheses and facts, a complicated medium through which the principles operate to order facts. In the case of Aristotle's biological studies, the accounts of the growth, decay and reproduction of plants and animals do not reach to the problems of heredity and genetics, or even to a unified classificatory system. In this case as in many other bouts in the anti-Aristotelian war of the last five centuries the battle was precipitated by the identification of an article of faith with a proposition of metaphysics and the application of the result to the findings of a new or renovated science. The creation of "kinds" in the Biblical account of creation was identified with the species of Aristotle's logic, and the result, "the creation of fixed species" is assumed in the classificatory scheme of Linnaeus in the eighteenth century. Thus Darwin is supposed to have upset Aristotle, whose metaphysics does not allow for divine creation but who did know about the continuum of forms that arises from the infinite of division and possible infinite variations which are the crucial assumptions in Darwin's evolutionary theory.

There seems to be nothing in Aristotle that would contradict the theory of evolution, although the minds of the two men never actually meet either in crucial data or opposing hypotheses. It is in the gap between them that the battle has raged in terms of the anomalous forms in classification, or the missing links in evolution. Each of the supposed giant antagonists would recognize the hypothetical status of both classificatory schemes and evolutionary series and the arguments that flow from each; many times their descendants with an infusion of stale theology have failed to keep the speculative distance which is required in the art of hypothesis. Consequently evolution, like Galileo's theory of falling bodies and the Copernican theory of planetary motions, has become a *cause célèbre* against ancient science.

The empirical method which has here been presented and illustrated in the work of two great empirical minds of the nineteenth century, Faraday and Darwin, consists in the use of principle and hypothesis in the discovery and ordering of facts. Sometimes the facts seem literally manufactured, made by the operations of human hands and instruments; sometimes they are selected, sifted, and arranged by the exploring and constructing imagination. But while the use of hypothesis and principle recedes into operations and rules that simultaneously take the abstractions for granted and also deny their existence, the emphasis and weighting of the voluminous, complicated, and ambiguous stuff of science gradually shift the balance until the positivist is correct in reporting that the method of science is simply the description of data. Hypotheses and principles are verbal symbols, some of them with hard factual foundations and others merely parading the airy fictions of word-magic.

Psychology

For the last twenty years the school of methodology that calls itself neopositivist and at present recognizes its critical leader in Carnap has trained itself to practice a rigorous discipline in the use of terms and symbols. They seek clarity and economy in scientific discourse, virtues which they find in the techniques of mathematical logic, and following the skeptical and ascetic style of Bertrand Russell, they practice what they preach. Somewhat naïvely, as it would seem, but realistically, as they would claim, they accept the facts of observation without much attention to the problems of surface and artifice that the long tradition of science has found in appearances. They recognize the function of hypotheses in the recording and ordering of the facts and assign this function to the valid and dependable use of symbols. In principles they find three uses or kinds. Some so-called principles are the rules of grammar which are employed in the formulation of hypothetical and factual propositions, some rules of a special form express equivalences or tautologies, and these they say are most often mistaken for metaphysical statements. The great majority of supposed principles are, however, nonsense parading as profundity.

This school represents the most successful protestant holding operation against the subtle and steady drift of the decayed dogmas of tradition into modern science. This is a protest not merely against ancient Platonism and Aristotelianism but also against the mixed metaphysics and cosmology of Renaissance science. It takes Francis Bacon, David Hume, and Immanuel Kant at their word and tries to keep winged symbols fully grounded to fact. But the speculative pressures, both from winds of doctrine and from factual ground swells, are considerable, and there are not many followers in the strait and narrow way. The pressures show themselves in many ways, but the most obvious and massive manifestation is in a tendency to explain away the difficulties in the position by referring the whole symbolic structure, its valid as well as its tautologous and fictional uses, to psychology. As Galileo and Descartes had referred secondary qualities, those that do not easily reduce themselves to quantitative elements, to the mind or the psyche, so these contemporary critics tend to relegate the puzzling by-products of their sorting to the nonexistent or mystically accepted human soul.

In the development of any method, in passing from partial to full application, there are of course anomalies, some arising from errors in observation, some from failures in skill and technique, some from uncriticized imagination and some from pushing logic to an extreme. The radical empirical method imposes a drastic discipline, and for more than a century there have been many anomalies of all sorts. Modern psychology is said to have originated in an early discovery at the Greenwich astronomical observatory of a systematic error in the observation of the midnight star passing the meridian and thereby starting the astronomical day. The secret of the Greenwich error was found to consist in the reaction time of the observer who pressed a telegraphic key some important fraction of a second after the transit of the star, and this in turn raised the question of the conduction of nerve currents and the formation of images in the mind.[30] This opened to investigation all the sensory illusions in which psychics and physics, which were supposed to operate in some pre-established harmony, showed systematic discrepancies and posed the general problems of mind and body. The positivists, in order to preserve empirical physics, had to become psychologists.

In this psychological salvaging operation there were several lines to take. A great deal of mistaken fact came from the persistence of folklore, popular or learned, and could be charged to superstition and the tyrannies of priests and kings; some more of it might come from earlier or recent stages in historical development in which institutions and professional disciplines had failed to disappear when they had lost their original function; still more came from social and self-deception, the symbolic satisfaction of outdated or perverse desires. All these are reminiscent of Bacon's Idols, but they were no longer assumed to be movable habits and customs; they were discovered as organic parts of the human psyche preventing and hiding the means of their cure. Latterly it has seemed that what Kant would call the kingdom of ends, the system of goods, has been handed over as a troublesome human obsession to be cured by a psychological theory of value.

Midway in the process of purging the illusions, Ernst Mach launched a countermovement within the empirical positivistic school. A book called *The Science of Mechanics* showed how physics could be revised and improved by stating all its laws as the description of sensation, and then he wrote another book called *The Analysis of Sensation* in which he showed how psychology could master its rather chaotic and anomalous subject matter by stating its laws as an alternative description of sensation. Physics and psychology had a common material, the data of

[30] The reference here is to the discovery by the Royal Astronomer Nevil Maskelyne in 1795 and 1796 that his assistant David Kinnebrook was recording the transits a half second or more later than Maskelyne thought he should. "Therefore," Maskelyne concludes, "though with great reluctance . . . I parted with him." An account of the incident and its sequel in astronomy is given by R. L. Duncombe, "The Personal Equation in Astronomy," *Popular Astronomy,* LIII (1945), 2–13, 63–76, 110–21.

sensation, and their two diverse analyses and descriptions, properly ordered, might possibly then be correlated. Mach's work set these two parallel projects of investigation for the positivist school, and it had wide effects on both sides of the common field of observation, human experience.

It has been said that modern psychology began by losing its soul, and that it then proceeded to lose its mind. But this, like many witty remarks, contains two truths. The investigations of the psychic phenomena, human experience referred to the individual, followed David Hume and others in failing to find the soul or the ego, but they did find in experience itself not only the drift of sensations, images, emotions and thoughts, but also the highly reflective powers of consciousness.

In ancient, medieval and renaissance psychology the essential property of the rational soul was its reflective power. All parts of the soul—sensation, memory, and the intellect—in their separate and related ways reflected the world. Perhaps the most concentrated expression of this is in Aristotle's brief proposition that the intellect is the place of the forms, the place of all possible forms.[31] Much of modern psychology is discovering or rediscovering that this reflective property is immanent in all psychic operations. A very modern expression of it is that the human psyche consists in the power to deal with symbols.

But for modern psychology this reflective fact takes on a queer significance. It seems to mean that man has to see nature and himself in a mirror which is his own consciousness. As C. S. Peirce has put it, man has a glassy essence.[32] The actual fact is more complicated: in order to know, man has to multiply his view of the world and himself in many mirrors. His view of the world is deeply and unendingly symbolic, and this at least suggests that the symbols of the middle ages and of renaissance algebra, far from being banished, are of the stuff of our

[31] Aristotle, *De anima,* 429a26–28; see also 432a2.

[32] In Peirce's essay "Man's Glassy Essence," *The Monist,* October, 1892, reprinted in *Chance, Love, and Logic* (New York: George Braziller, 1956), pp. 238–66, a possible meaning for the title is to be found in the conclusion that "a person is only a particular kind of general idea"; according to Peirce, "all that is necessary . . . to the existence of a person is that the feelings out of which he is constructed should be in close enough connection to influence one another." The title is taken, the reader will probably know, from Isabella's speech in *Measure for Measure,* II, ii, 116–21:

> . . . but man, proud man,
> Drest in a little brief authority,
> Most ignorant of what he's most assured,
> His glassy essence, like an angry ape
> Plays such fantastic tricks before high heaven
> As make the angels weep.

experience. The data by which we know and judge all that we do know, the raw material of the empirical method, the abstract content of hypothesis and principle, and the imaginative medium in which these float are essential constituents of the psyche, and therefore are symbolical and instrumental. The ancient and traditional doctrine concerning this is that the rational soul pervades all of the functions of the rational animal—all the stuff of experience that partakes of the intellectual operations—and therefore reflects and intends natural things; it is the means of knowledge. But the rational psyche can also reflect itself, it can know itself in its own operations, and it can make itself and its operations the subject of scientific study. Mach's proposal of the two sciences, physics and psychology, is a positivistic empirical restatement of the ancient doctrine without the benefit of the hypothesis of the reflective rational soul.

A less complicated statement of the central point has more recently been made by Bertrand Russell. He postulates as the initial data of experience neutral entities, things that are neither psychic nor physical. These data referred to physics are aspects of objects; referred to psychology they are items in a perspective. The system of aspects is nature; the system of the perspectives is mind.[33] A slight familiarity with the elements of projective geometry suggests a possible relation between mind and body, but for the uninitiated Lewis Carroll's Alice books will serve the same purpose: *Alice in Wonderland* shows what happens when the neutral entities are referred to physics, and *Alice Through the Looking Glass* shows what happens when the neutral entities are referred to psychology. In an almost incredible way these two books have become the modern man's *Divine Comedy*. Mathematical physics is surely a wonderland, and psychology is the world of the looking glass. Any mathematician who can write prose is perhaps our best Virgil in wonderland; our guides in the world of the looking glass must correspondingly be William James and Sigmund Freud. James is the interpreter of the surfaces and Freud leads into the depths.

[33] For expositions of this view, see Russell, *Our Knowledge of the External World* (New York: W. W. Norton & Co., 1929), Chapters III and IV, and *Mysticism and Logic* (London: George Allen & Unwin Ltd., 1959), pp. 139ff., 158ff.

William James

William James had some medical training and some working experience as a painter and he therefore had some skill in observing the operations of the human soul both in himself and in other people. But unlimited exploration in this land of reflected light was also a tradition in his family; his father and his grandfather cultivated a native mysticism in themselves. In the middle ground between religion and science there was also room for philosophical inquiries. All of these went into *Principles of Psychology* and came out of it criticized and renovated; also the psychology of the previous hundred years was processed by a fresh, critical, and inventive mind so that *Principles of Psychology* has been the starting point for many new lines of investigation. It has also had its influence on schools of philosophy and literature.

A great deal of James's influence in fiction and literary criticism is direct: his own literary style has been imitated. It is true that he wrote psychology like a novelist, as his brother, Henry James, wrote fiction like a psychologist. But the style was a novel invention arising from a unique blend in himself of the two mental traits that he found dividing the human race. He was tough minded in that he had a robust respect for morals, logic, and facts, but his respect for facts included a sensitive response to the variety, ambiguity, and wild noncomformity of experience to abstract laws and rules. He found the endless antinomy of determinism and freedom a ubiquitous source of wit and irony. His criticism of the introspective and brass-instrument schools of psychology is illumined by this kind of fun. He had much to do with the revival of Yankee humor as an antidote to the pedantry of German scientific scholarship, in which the spade work of psychology had been done. He was well fitted to start out from the limbo of common sense to search for the reflections of physics in the labyrinth of the human soul.

One of his most telling and definitive figures of speech is the description of experience as the stream of consciousness. This is a typical reaction to the British and German analysis of experience into atomic elements in sensation and their bricklike articulation into the higher mental structures. There are discernible items and structures in consciousness, but they float or swim in a flowing field in which there is a central current of activity, whirlpools and backwaters, shoals and depths, and a constantly changing course that reminds one of Mark Twain's Mississippi. This was at once a break from the discrete and structural descriptions that preceded it and a positive affirmation of the dynamic and functional view of mental life, and it was backed by shrewd and delicate observation. It amounted to a redefinition of all psychological problems.

James was not alone in this revolt against the blockheadedness of the structural and mechanical modes of analysis that had been imitated from physics. There had been a French ferment and conspiracy against the measuring rods and clocks that had made the physical world into a three-dimensional game of chess. James simply moved with Alice and her cat into the looking-glass house in which he found many more things than mathematics and physical theories of time had hitherto dreamed of. He noted how memory and imagination not only reverse the irreversible and live backwards, but also direct the course of the current and tear away the muddy banks of the stream of consciousness. Objects that seem to have been torn loose from the banks of the stream higher up and to be passively floating begin to swim of their own energy and sense of direction and purpose. What nature had lost in the human effort to understand it, its animateness, is restored in the life of the soul, and in addition there are struggle and conflict, variety and adaptation, that put Darwin's system of nature in a shadow. James was careful not to revive the theories of the soul that had been parodied and stultified by the war between religion and science, but he recovered the original meaning that the word has, the principle of life.

But as in Lewis Carroll, the looking-glass house does not always transform and distort the physical world. Some symmetries and some other patterns go over from physics to psychology without change. It is not certain that all such patterns in James's version are validly transferred without change. There is considerable dated and now outdated neurology imitated from a semimechanical model that he packed up and took with him on his hunting trip. He accepted, to be sure with some skeptical criticism, the rather vague hypothesis that psychical function was somehow carried by the nervous system, and that theories of seeing, hearing, smelling, tasting, and touching as well as some other "higher processes" ought to be explained by the chemistry and physics of the nervous system.

But even these glances back into wonderland served as searching working hypotheses that sharpened his vision for facts which often admittedly refuted the hypothesis. Of such cases perhaps the James-Lange theory of emotion is the most interesting. James proposed that our awareness of emotion is a psychic recognition of automatic physiological reactions to stimuli on the part of the autonomic nervous system, that we are sorry because we weep, and that we are frightened because we run away. James is here accepting Darwin's account in *The Expression of Emotions in Animals* and letting the animals live as if independent inmates of the looking-glass house of a man's soul. It is a witty and illuminating episode for him, but it is not well digested and assimilated to the stream of consciousness. The more radical magic of Freud is needed to transform physics to psychics.

There are many such incomplete translations in James, most of them approaching as a limit the ideomotor theory of the will, which is probably his most favorite theme. All the elements of the psyche in the dynamic, functional aspect of the stream of consciousness have an active phase, a tendency to move as well as to reflect. Ideas are not only rules of operation, they are causes of motion in themselves and in other things. Thus impulses, instincts, habits, innate or acquired drives, are elementary or composite parts of the will. But the series of organized means to ends, or the instruments of the fulfillment of such tendencies, are not always chosen with discrimination from the two worlds, the psychical and the physical. A piece of physiological, chemical, or physical machinery is found coupled with carefully observed and described psychic functions and presented with witty and imaginative literary exposition whose charm persuades without the need of rigorous argument. Many a contemporary realistic novel written with less art on James's psychological themes leaves the reader guessing, as Lewis Carroll does at the end of *Through the Looking-Glass,* whose drama this was.

The spell that James weaves has carried far beyond the direct influence of *Principles of Psychology*. It has founded and propagated a new school of philosophical criticism, one that is now known all over the world as a strictly American philosophy—pragmatism. James borrowed its main doctrine from an essay of C. S. Peirce, *How to Make Our Ideas Clear*. Peirce is there expounding what is now called operationalism, the doctrine that any abstract idea contains, as it were, a rule of human action or natural change, and that the ambiguity and confusion that abstractions occasion in human thinking can be clarified by the explicit statement of the rule and its application to concrete fact in experience. This theory or method for clarification of ideas was fused, if not confused, in James's mind with his own theory of the dynamic property of ideas, and he further identified it with the truth of ideas. If a rule that is formulated for an idea works in practical application, the idea is true. In the old theory of the soul, the psyche has two essential powers, the power to reflect, which in man amounts to the power to know, and the power to move. In its upper limits this distinction of powers led to the ultimate speculative puzzle as to whether the theoretical or the practical intellect was superior and controlling. James is facing this high speculative question on the empirical level and strongly defending the position that the practical will is dominant and that therefore in general application the pragmatic test supersedes the theoretical.

This doctrine in the minds of James's followers has had many ramifications and consequences. Notably in the wise and commonsense mind of John Dewey it has become a philosophy of science attuned to Anglo-Saxon countries where the weight of social responsibilities for the consequences of the industrial revolution has needed an open-minded explorative consideration in the midstream of action. It is doubtful that it would have had such ready and widespread acceptance if the shift from the laboratory and observatory to the marketplace and the parliament in the last two hundred years had not added its momentum to the psychology of William James. It is notable that C. S. Peirce, with the strong theoretical bent of the laboratory, wrote his essay on the clarification of ideas for the *Popular Scientific Monthly*. It is doubtful if he would have gone all the way with James and Dewey in identifying the true with the good, or that he would have been happy to have James borrow the word *pragmatism* from his pragmatism.

Freud

WHILE James was making his comparatively superficial examination of the house of the looking glass and becoming conversant with its lively inhabitants, another explorer, also with a medical training, was making deeper observations in the larger and more puzzling world surrounding the house. Sigmund Freud recognized all that James had found and implicitly accepted it; it was in the current of European empirical thought. His view of psychological phenomena was also dynamic and functional, but it seemed to him that in the stream of consciousness there must be deeper and more rigorous laws governing the psychical world, independent of the laws of the physical world. By a combination of two techniques, hypnosis and the interpretation of dreams, he broke through the magic mirror of the veil of sleep and in no time at all found back of the ego, which sensed, remembered, imagined and willed, the realm of the unconscious which truly had its own laws and manners.

Behind and below James's spontaneities, drives, and purposes in the foreground of consciousness, and also beyond a kind of vestibule where recalled memories might be met on occasion, there was a somewhat shadowy underworld in which originals of the bright conscious symbols moved ponderously and powerfully to control the surface navigation on the stream of consciousness. Freud's clinic in psychopathology made him dive beneath the surface of troubled waters, and he came up to report that all the surface phenomena have hidden causes. The psyche was not so much a stream as a large primitive organism, amoebalike, whose interior was a field of forces only partially manifest in the data of consciousness. These forces were manifest to the diver in dreams, word associations, or those artificially induced dreams that the hypnotist could observe. The means of observation in this realm were not telescopes or microscopes; they were symbols which reflected recognizable things, not so much in the wonderland of physics, but rather the immediate social environment of the individual, particularly the members of his family. The ordinary, natural world had bombarded the individual psyche with stimuli so many and so powerful that the primitive organism had become tangled internally and calloused over externally, leaving only the focus of consciousness as its medium of intercourse. Through this window, or highly selective looking glass, images of parents, relatives and sweethearts had passed and now inhabited the interior. Not much more of society than the immediate family and their associates had so passed and were then reflected in dream symbols and word associations, but this much with what was known of the tribe, the primitive family, provided the clue to whatever else was to be found there.

Freud was apparently not much interested in the study of what is now called social psychology, the explanation of the individual in terms of his social environment and human relations, but he brought his clinical findings that partially reflect these into a focus by borrowing from his Gymnasium training. The family relations revealed and eloquently traced in Aeschylus and Sophocles, with a few additional points from Greek mythology, supply him with the terms and relations that his analysis demands. These relations are relations of love, and in the dramatic context of the family with its regular crucial episodes of birth, death, infancy, childhood, adolescence, and maturity, the sexual grammar of love determines the distribution of a general psychic energy which is called libido. In the infant this psychic energy, like the One of Plotinus, emanates and overflows into all possible forms, but soon the limited possibilities in the external life of the family and the watchful eye of the conscious mind select and arrange the actual channels and course of the stream of life.

At each stage some possibilities are realized in the fulfillment of internal wishes in the external social world. Some are only partially realized in the rituals of day-dreams and night-dreams, and still others, the great majority, are suppressed without the individual's awareness. As manners, customs, and laws in external life make life possible and secure by prohibiting a great deal of the external expression of the libido, so the psyche sets up its own device of self-censorship, a sovereign symbol which sorts and dispatches the wishes, facilitating some, inhibiting others, and appeasing and conspiring with others to give them substitute symbolic satisfactions. Although most of the operations of the censor, or the superego, are subconscious, and only a few appear in conscience, the results of suppression are often manifest to the psychoanalyst in the hysterical symptoms of psychogenic disease or the commonplace aberrations of daily behavior. It is the function of the psychoanalyst to discover and to correct the apparently arbitrary and erroneous acts of the censor, and to make it possible for the conscious patient to understand by seeing the real nature of his desires.

Although the internal drama of the family symbols reflects only the immediate external environment, the censor gives some shadowy composite picture of the institutional structure of society. Unfortunately the outlines of this projection are confused and generalized. Freud tends rather defiantly and naïvely to see the artifices of society, the laws of government, and the rituals of religion as harsh, primitive, tyrannical inhibitions and even punishments of the free expression of the psychic life, although he recognizes freedom in sublimation.

Some of this foreshortening of the projection in the looking-glass world is due to the great difficulties of the art of psychoanalysis, and with an awareness of the incompleteness of his new science, Freud very tentatively proposes a metapsychology, a speculative completion of his observations, to suggest the lines of further investigation. The leading idea in this is psychic economy, the notion that there are systematic oppositions in the fundamental forms of psychic energy and that the best the psychic organism can do is to maintain some sort of dynamic equilibrium between the conflicting forces. There are four such fundamental drives, all of them together comprehending the full field of energy in the psyche. The life and death instincts comprehend together the separate wishes and their balancing of each other. The reality principle comes into play when the resolution of the life and death instincts successfully reflects symbolically the external world and connects the wish with its object. The pleasure principle operates to support a merely symbolic resolution of the internal conflict. Needless to say they all come into play in all psychic life.

But this highly abstract, speculative metapsychology was a small convex mirror hung in a distant vista of the looking-glass world as a kind of landmark and indicator of the road of investigation which would have to be travelled and worked. An early dissident from the Freudian school of psychoanalysis was not happy with the exclusive predominance of the family in the foreground and decided to explore another avenue. Freud had used the contemporary bourgeois family as the model for the projection of symbols into the psyche, and when that, with the interpretive help of Sophocles and Aeschylus, had delivered its pattern of social relations, he jumped back to the primitive prehistoric tribe. Carl Jung saw the resulting interval as the possibly rich repository of race memory into which many other human social patterns might have been projected. In this way many of the shadowy generalities of the customs and institutions which Freud had identified and possibly confused in the condensed image of the superego might be clarified and articulated.

For instance, many of the sexual patterns of the Freudian complexes and their sublimations are to be found in myths and rituals of the various religions of the world, and in that context appear less important and basic than in the tight patterns of Greek tragedy. Alternative patterns were found to be pervasive and recurrent in the interpretations of hypnosis and dreams and to have their utility in psychotherapy. One need not accept the hypothesis of race memory, but may be grateful for the notion as a leading principle by which more of society can be projected and found in the psyche. One result of it would also be that psychoanalysis, or analytic psychology, as Jung calls it, might become social in the practice of its techniques. Members of the Jungian group, by undertaking a project in organized research on the common heritage of ancient historic symbols, have become better observers of contemporary institutions.

Modern psychology in the last two hundred years, rising as it did from the rigors and purges of empirical science, has tended to run in schools, groups splitting from groups or starting up at some new glimmer of light. They have been swimming in the depths of the psyche or navigating on its surface. The vision is dark and fragmentary, but there is no doubt that it has become a new science, and its vagaries and its episodic character are merely signs of a young science operating in the atmosphere of the empirical method. Part of the apparent bizarre and occult atmosphere is due to the early use of the methods of mesmerism, phrenology, and animal magnetism in the procedures of investigation. This has persisted from the early days of brass-instrument psychology that James makes fun of to the modern techniques of psychoanalysis and the electronic amplification of nerve currents. But these methods are to be associated with the peculiar circumstances that threw up the new data to be salvaged and ordered, the discovery of the radical systematic error in the observation of facts. As the techniques of experimentation and measurement became complicated and precise, the increased selectivity of the technique left more and more of the original stuff of science outside its increasingly strict powers. To meet these remnants of appearance, old methods were revived and new ones extemporized. It is not surprising that there have been societies for psychic research that have applied new and old techniques to the occult science of necromancy to bring the inveterately speculative beliefs in the survival of the soul after death into scientific respectability. The figure of the looking glass has here been used to hold the fragments of psychology together for a minimal single view, and at the same time to hold the ragged fringes of the vision from interfering with the focus.

But there is something familiar in this material to anyone conversant with the scientific books of the West. The top of the scientific hierarchy was lost in the clouds of speculation at the end of the Middle Ages. In the misty heights there was a possible meeting of two great scientific traditions; mathematics and love met in the mystic beatific vision in which all things could be known in their true being. Mathematics has done very well in absorbing the natural sciences and reaching new speculative heights. The philosophy of love, the kingdom of ends, and the theory of value have not done so well. In fact in the middle of the nineteenth century they had all but disappeared except as they figured in the somewhat empty rhetoric of the so-called social sciences. In almost a Freudian manner they had been suppressed, with only poets and dreamers protesting. On the other hand the empirical movement in science was approaching the pragmatic criterion as the justification of its curiosity.

It was in this circumstance that Psyche, as in the myth, was invoked to sort the seeds of good and evil. She came with some memories and some guile. Love was her dream and her only wisdom. She at first gathered the seeds dropped by the master mills of science, and then with many mirrors given her by Ernst Mach and the new converts to her cult, the psychologists, she rebuilt her court of love out of the shadows and reflections that the laboratory could not use. In place of the soaring pinnacles and spires of traditional theological and mystical architecture she has followed the threads of time through the labyrinths of memory, dissolved the walls of conventional time, and recovered by hypnosis and dream-reading her own buried treasure of forgotten symbols. Psyche's task is not yet finished, but the many followers in her staff of research are catching up with the march of the natural sciences.

One sign that this is happening is to be noted in the new time structures that are being imported into physics. The revolution in physics that resulted from Einstein's reformulation of the principle of relativity opened the door for many new inventions in time schemes. Mathematics from the time of Newton and the use of fluxions has been able to incorporate time into the laws of nature. Clocks have been invented to embody clarified fragments of duration. But as Kant said, time is the schema of the internal sense, the essential fabric of memory. Physics relies for its correlations more and more upon the analytic or algebraic method in mathematics; algebra becomes more and more arithmetical in its interpretation of its symbols; and arithmetic mines in the quarry of memory for its elementary schemes of counting. Perhaps Bergson in *The Immediate Data of Consciousness, Matter and Memory,* and *Duration and Simultaneity,* has served the purposes of Psyche most. The deep essence of time is manifested in the intuitions of duration, which are prior to the intuitions of time and space as Kant had formulated them. Duration is deeper than the flow of experience, in fact it is the spring or fountain from which the stream of consciousness rises. It is the common and continuous source from which all schemes of time are derived by abstraction, and from which we may always expect revolutionary novelties in the logical constructs of science. The most recent critical speculations concerning the paradoxes of the expanding universe and the displacement of lines in the spectrum which seems to be evidence for it have turned up a tentative solution or redefinition which involves such a novelty. It may be that the so-called modern Western obsession with time will give a new meaning to the ancient formula that the soul is the form of the body—of all bodies.[34] The unification of science may wait upon some such end to its present paradoxology.

[34] See Aristotle, *De anima* II, Chapter 1.

Part III. The Unity of Science

THERE is a legend concerning Saint Thomas Aquinas and his *Summa Theologica*. He is reported to have expressed his discouragement in a late stage of the work by saying to himself, "It is all straw." Thereupon the image of Christ appeared to him and said, "Thou hast done well by Me, Thomas." This legend bears pondering because it expresses rather fully the nature of the whole human intellectual enterprise. Quite apart from what may seem to the uninitiated as dogmatic special pleading for a revealed truth, there is in the *Summa* an effort to unify knowledge, which does not dodge difficulty nor flinch at paradox, faces evidences of faith as well as of fact, and follows the ingenuity and continuity of an imaginative intellect to the bitter end. The happier, freer, and perhaps looser and wiser version, Dante's *Divine Comedy,* is a tribute to the Angelic Doctor, confirming by reflected light the reality of the Thomistic vision. But for Thomas himself his masterpiece was straw. In theological terms Christ's response is the oracular utterance of the subject matter itself, the Logos, which again in theological terms is the subject matter that all science seeks and approaches as its limit or final end. The legend says briefly that the human demand for the unity of science is irrepressible, stopped neither by partial success nor by failure.

The typical Oriental expression of the demand is in the many sets of fourfold ways to wisdom, and these have probably entered Western thought through the Arabs as the methods for improving the human understanding, but the first explicit dialectical treatment of the demand is probably in Plato's *Charmides,* in which Socrates makes the hero, a healthy man with a headache, listen to the diagnosis of his trouble as the unsatisfied demand for a science of the sciences. The theme has been elaborated many times in the Western tradition, but the Socratic oracular conclusion, "I know that I do not know," has not been improved upon. St. Augustine and Kant repeat it with its ironic overtones; modern man repeats it with literal dogmatism. And yet the search and the demand go on.

The demand for unity in science is not merely the arbitrary postulate of the philosopher nor the untutored wish of the common man who finds the doctors disagreeing. It is implied in every step of investigation

—the search for principles, the collection of data, the construction of hypotheses. In each of these modes there are moments or phases in which the inquiry demands division and isolation of parts, but the aim of these most particularizing drives is the finding and holding of a part of a whole which is one, and the other moments or phases are obviously the combination of many into one view. The secret of the partial successes has for a long time been recognized as the discovery—or construction—and use of abstract ideas. Depending on the mode of discovery or construction and the purposes for which abstractions are used, they are called by different names: ideas, forms, concepts, universals, fictions. They reflect and order particulars, they combine to make discourse, and they raise the radical question of their truth and falsity. They are universally recognized as the good or bad work of the intellect. These foci of unity are the deepest concern in science.

A great deal of Aristotle's pioneering fatherhood of the sciences was occupied with these tools of the intellect, their origin in the process of abstraction, their functions as predicates in the propositions of discourse, and the reflective powers by which they become the medium of knowledge. Plato had surveyed the field and shown its possible complications. Aristotle cut it at the joints and defined its original regions, and in so doing discerned the arts by which the regions might be cultivated and the skills and capacities that these arts required in the intellect. These capacities and skills have a wide range of exercise, some of them theoretical, as for instance when they appear in intuitive induction as intellectual intuition of principles and the deductive ordering of hypotheses; some of them practical, as in the making of useful things and their use in the conduct of life; and finally, in the apprehension of the "highest things" under which all other things are subsumed. These capacities are exercised in all human affairs, but for the purposes of science they become organized in the intellectual or liberal arts, the arts for dealing with universals.

In Aristotle's *Organon* these arts are sorted and allocated to the various sciences and through them to their application in various human affairs, but for the followers of Aristotle and Plato they become a circle of disciplines, an encyclopedia, for the instruction of children and recruits to the incipient learned professions, particularly to medicine and law. They are formulated in two groups, the three-way or trivium, and the four-way or quadrivium. The trivium, grammar, rhetoric, and logic, deals with the makings and understandings of verbal symbols in language; the quadrivium, arithmetic, geometry, music, and astronomy, deals with the symbols of mathematics. This division and corresponding divisions within each group are based upon the fact that verbal and mathematical signs are artificial products of conventional usage. Grammar, arithmetic, and geometry make, use, and analyse the signs and their combinations; logic and astronomy attend to the abstract meanings that the signs carry; rhetoric and music deal with the applications to concrete things. It is recognized that these intellectual arts are basic to all human thinking, learning, and knowing. A child becomes free when he acquires the skill and knows the reasons connected with the artful use of the symbols of universals in language and mathematics.

The tradition of liberal education is based on this foundation and it has grown by the use of the books in literature and science that are exemplary products of the practice of these arts, books that many times in history have been selected and listed, and sometimes published for educational use. This tradition and even the terms of its articulate formulation continued from the time of the Greeks to the early part of the nineteenth century without break. It could be argued that the recent revival of semantics, the science and art of signs, is a part of this tradition, although its followers would deny it. During the unbroken period of the educational tradition there is a very remarkable continuous change in the understanding and use of the liberal arts. Universals, or abstractions, have many dimensions and handles. During the early Greek period, although there were many epicycles, there was one large cycle of change, in which at first rhetoric and grammar were used for the purpose of isolating universals, then grammar and logic were used for the purpose of persuasion in political rhetoric, and finally logic and rhetoric were used for the description and collecting of facts. One can see in the account of a similar shift in scientific methods, as described in the earlier parts of this essay, a large cycle in which the earlier Greek and medieval scientists were seeking principles, using grammar and rhetoric for arriving at higher and higher abstractions. In the Renaissance the logic of principle and the grammar of fact were subordinated to the algebraic rhetoric of the great hypotheses. Finally, empiricism in the modern period can be described as the instrumental use of logic and rhetoric to uncover and order facts, according to the grammar and arithmetic of nature.

There is no doubt that the liberal arts have been lively in their historic career. A complete history of them from the Greeks to the present might throw some light on the mysterious disappearance and reappearance of certain themes in the discourse and practice of scientific thought. Needless to say, this essay provides only episodic illustration of such a history, which could be accomplished only by a team of learned men. But there is one persistent issue which is a matter of common knowledge and can be reported here. It is perhaps the most discussed question in human history, the nature of universals. The classical statement of the question was made in the *Introduction to Aristotle's Categories* by

Porphyry, the disciple of Plotinus, in the third century A.D. "Genus and species" is the appropriate reference to universals:

> . . . I shall refuse to say whether genus and species are subsistent or lie only in the bare intellect, whether, if subsistent, they are corporeal or incorporeal, and whether separated from sensible things or subsisting in them or about them. That business is very profound and requires a more extensive investigation. Nevertheless, I shall try to show you how the ancients and especially the Peripatetics treated this and the other proposed subjects dialectically.

Porphyry's refusal to answer the nest of questions and yet his willingness to give some account of their dialectical treatment are parts of the issue. He is aware of the two ways in which the issue had been met up to his time, and, we can add, the two ways it has been met up to now. The answer given either way helps to locate the probably incurable trouble concerning the unity of science.

The central question, which Porphyry does not ask because he has already accepted an affirmative answer, is: Do universals exist? The explicit answering of that question would be the "very profound" business which lies back of the dialectical treatment to which he alludes. It is a metaphysical question which any serious specialist in logic or pure mathematics cannot avoid asking. Asking it, however, does not lead to a simple answer. The simple, general, affirmative answer demands the extensive investigation of the various kinds of existence or the degrees of being that universals can have. It also involves the sorting of evidence that comes from a critical scrutiny of the practice of the liberal arts.

The skillful logician or mathematician finds the evidence in his own practice overwhelmingly affirmative. If these specialists allow themselves the freedom or indulgence of metaphysical speculation, they affirm the existence of universals and draw the inferences from that affirmation that determine a realism in metaphysics. The varieties of realism are many, depending on the answers to the secondary questions which Porphyry asks. Universals subsist in the human intellect, in the Divine intellect, or in natural things, in two of these places, or in all three. In the human mind the universals fall into a hierarchy of ideas; in nature they make a system of forms; and in God's mind they are sublimated into a deep simplicity.

The trouble with the boldness of the logician or the mathematician who lets himself go metaphysically is discovered in the infinite discrepancy that he immediately discovers between his speculative vision and the fragmentary and stumbling practice of his art, no matter how far his skill has taken him. His *hybris* in scaling the heights induces a fall to some level from which he can view the pattern laid up in heaven and on which he can set his house in order. The tragedy can be seen in comic form in the encounter between the youthful Socrates and the aged Parmenides as Plato describes it in the dialogue, *Parmenides*. It is a tragedy re-enacted by every generation of metaphysicians; they are always friends of the ideas recruited from the most skilled of the liberal artists.

Some liberal artists who specialize in the more constructive arts of grammar, arithmetic, and geometry give the negative answer to the central question. They find ample evidence of the fictional character of so-called universals in their own invention, construction, and manipulation of signs. Signs that carry general significance, like natural objects, move or are moved in space and time. The things that the realists call universals are corporeal, are bodies, whose operations do the work of the abstractions. Signs are names, moving names, of the bodies which they imitate. Those who believe thus are the nominalists, the friends of bodies, whose boldness and whose troubles are classically described in Plato's *Cratylus*. Their chief trouble is the demand placed on them for a multitude of novelties and inventions in their practice of the liberal arts and science. The empirical materialist must catch and anchor his mythopoetic imagination to solid body with many mechanical gadgets, some of which he has to admit are shadows and reflections in rapidly flowing water. The unity of science for the nominalists is in the temporal flow of the stream of experience and in the infinite drift of the atoms in the void.

But these are the extremes in the unending battle of the liberal arts. The realists can be moderate when they descend to the domestic level of application of their regulative vision, finding universals embedded in bodies and carried by images. The nominalist usually accepts forms as relations between parts of bodies, and the great majority of metaphysical speculations succeed in some working articulation of form and matter, the unity of nature finding itself reflected in the structure of ideas. The art of rhetoric, which has had its origin and longest practice in courts of law, brings the extremes together for an adjudication of the profound business, and although the meeting of minds is often an empty formal affair, the substantial result may be deep and comparatively lasting in its unifying effect. The case of idea against body is capable of analysis by the distinction and redefinition of terms, so that pure form and prime matter divide in order to unify everything in the universe. The case always recurs. The celebrated occasions in the dialogues of Plato and the treatises of Aristotle in the ancient world, the *Summas* of the scholastics culminating in the *Summa Theologica* of Thomas in the medieval world, and the *Critiques* of Kant in the modern world have marked the real crises in the quest for unity and have moved science into new, if not better, positions.

But the art of rhetoric in its familiar manifestation as the process of persuasion has two ends. At its best it brings truth to men, but much of its effort is spent in moving men toward their apparent goods. It does this by using all the furniture of heaven and earth. Nothing is alien to the art of rhetoric. It therefore intensifies the demand for unity and is capable of the most versatile wiles and winning ways. It poses and dramatizes, it illustrates and argues, it spins allegories and lesser figures of speech. It is social and communicative between persons, but it also has, as Plato pointed out, a whole life inside a man. This internal rhetoric of self-persuasion is a conversation imitating and intensifying its more external manifestations. This internal rhetoric is often called dialectic, the Greek word for conversation. It was inevitable that Hegel, or someone else, should have seen this dialectical archetype projected into the world of history. The battles and alternations of aims in the liberal arts are dialectical, the adjudications of the schools of science and metaphysics are dialectical, the passage of political dynasties and civilizations is dialectical. History is the locus for finding the unity of thought in the dialectical mode. The fact that the irrepressible demand for unity has produced many schools of metaphysics appears to be a sign of systematic and unavoidable defeat, but for the dialectical historian these articulations of profound matters are eloquent, heroic voices in a great conversation, a conversation which is not less significant because the actors are very often unaware of their roles, and never completely aware of the plot that comprehends them in a gigantic drama.

Hegel's voice itself, whether it is heard in the internal dialectic of his *Science of Logic* or in the external dialectic of his *Philosophy of History* and *Phenomenology of Spirit,* is Stentorian, the voice of a man standing at a special moment of history, the time of the French Revolution, and seeing all that he knew of human history as one great drama. His eye reaches back into the shadows of distant time when there was no history, when neither the individual nor the group even asked the question of roles and plots, but in the absence of light, spirit manifested itself in movement. In the beginning twilight there was feeling, which in the succeeding periodic motions came to know itself. As if to match scenery with plot these beginnings are placed at the most distant spot of the earth's surface, China, and the action moves westward in time. The process consists in a series of emancipations from darkness to glimmers of light, each of which in turn becomes opaque and brings on a long period of self-enslavement, a kind of self-hypnosis generated by a static, partial light. Hegel believes that, although one swallow does not make a spring, one act of self-knowledge can make a renaissance; a hero can make a revolution, provided he comes to know himself as the embodiment of the spirit of his time. Such self-knowledge is the rebirth of spirit and manifests itself in a revolution in culture. Furthermore, such revolutions in culture are advances in truth and in freedom. The great hero is spirit itself, in which the heroic mind participates, and the plot of the drama is the struggle of the spirit to free itself.

Hegel made special studies of this process on various levels and phases, and these resulted in philosophies of art, of science, of law, and of civilizations. Each of these in turn became a voice in the great conversation. There have been many attempts before and since Hegel to put the science of history together. Most of them have followed a theory of cycles more or less closely, as the word "period," a circular path, shows.[1] Sometimes the cycles were parts of descending or ascending spirals, sometimes epicyclical episodes in a large revolution, and sometimes showed the wave-pattern of rise and decline as measured against an axis of reference, as in Augustine's *City of God*.

There has recently been an interest in collecting and reviewing these restudies of history. Vico, Buckle, Comte, Spengler, and Toynbee are familiar examples. In most of them there is an uncritical acceptance of some large pattern of combination for the fragments with which they deal, but there is even more an uncritical acceptance of the units which they assume to be the elements of the combination. The first volume of Toynbee's *Study of History* raises both of these critical questions, and the rest of his study shows the great difficulty of adequate answers with the enormous additional data that modern history has accumulated. On the whole, Toynbee accepts a wave theory of rise and decline for his unit civilizations and an axis of reference tilted toward progress. He has an unfortunate antipathy to abstract science and philosophical speculation which lets him ignore Hegel's powerful contribution to his problem of pattern and unit.

Where Toynbee tells a myth of challenge and response and pictures men struggling to master the natural or human environment in all their manifold mixtures of aids and threats, Hegel finds everywhere the deep antitheses of theory and action, and many distinct levels and phases on which contradictory ideas and opposing actions face each other. Any apparent unity, because it includes and comprehends oppositions, becomes a challenge to analysis, and any pair of opposites, because it arises out of an underlying unity, becomes a challenge to synthesis. Nothing is

[1] The Greek *periodos* is from *peri* ("round," "about") plus *hodos* ("way," "path"), and means "a going round," "the way round," "circumference," "circuit," as well as "a cycle or period of time."

understood until its diverse elements are placed in a functional pattern. This is as true of levels, phases, and elements within a civilization as between the larger patterns. Consequently, each civilization is a world, and one world civilization is related to another as a necessary opposite replacement in time. This appears to the ordinary reader, as it does to Toynbee, as rash and dogmatic speculation, but it is based on a modest and precise observation of the internal machinery of thought, which for Hegel carries the essential process of human history. Reason is of the essence of history and the external drama represents its struggles to recognize the consequences of its own operations; the freedom of the spirit is measured by the success of these efforts.

Man always steers between two illusions when he acts: the Scylla that tells him he is spontaneous and impulsive, and the Charybdis that makes him feel the pressure of necessity in his circumstances. In spite of these two deceptive mirrors, if he is wise, he chooses one among many alternatives, only to find, if he is honest, that the alternatives he intended to avoid turn up in the means, or the consequences, or the conditions of his action. He is caught in conflict and the need to compromise, and this is as true of selfish as of dutiful action. The simplest choice sets off a chain reaction of internal rhetoric or dialectic which through casuistry or rationalization, or both, results in the sacrifice of an original end and its replacement by another end which is general enough to save the concrete conditions. The classical studies of this universal human experience are in the Greek tragedies from which Hegel learned much of his dialectic; a contemporary study of it is in T. S. Eliot's *Murder in the Cathedral.* Hegel finds all rational human action shot through with fatal internal contradictions that force recognition, discovery, and acceptance of ever deeper and wider ends. For him reason is that mythical bird, the phoenix, which continually consumes itself by fire and rises from its own ashes stronger and wiser than it has been before. The special histories of the arts and sciences, as well as world history, which is best told in political terms, move through these ordeals by rational fire in which everything and nothing is lost because all is transfigured.

The dialectical process is obvious in the history of any hypothesis in science. A hypothesis is discerned in one or more instances of a phenomenon; it is formulated and generalized to indicate more instances some of which fit and some of which do not fit; it is revised and reformulated until it comes into conflict with other hypotheses and raises questions about principles. It is finally dropped altogether to give place to another hypothesis, or it is transformed and loses its original identity. In the case of systems which tend to become cosmologies, the inner contradictions emerge as antinomies that eat away the fundamental assumptions on which the sciences are built. Hegel found Kant's account of this process of frustration and his consequent strictures on all speculation intolerable, and it was this crisis in speculation that led him to the invention of his own dialectical method, a method to perpetuate all methods including itself.

As has been suggested, this method grows out of the subordination and use of grammar and logic for rhetorical purposes. Hegel's insights and skills belong to the art of rhetoric, in which the idea, the universal, is understood in its manifold relations to images and facts and to other ideas. The idea is not the object of thought, but rather the medium by which we know. Hegel's so-called logic is a kind of conjugation of all the possible relations that an idea can have, and his dialectic consists in shifting the sense and direction of knowing to all possible positions with respect to the origin and axis of reference. Every idea has a subjective and an objective phase or moment; it has an inside and outside reference; it is both abstract and concrete. An idea has a life like that of an organism, in which there is a beginning, middle, and end; there is a complex reciprocity of means and ends, and finally a growth and decay in time. Hegel himself corrects this analogy by pointing out that the organism has limits of comprehension and influence, but since an idea is a thing of the spirit, it knows no such limits and has a phoenix-like immortality. It is the soul or essence of history.

It is not surprising therefore to find that this super-artist in rhetoric should find politics the master art and the state the medium of the spirit in history. Following Rousseau, Hegel finds that a man, an individual spirit, realizes his own individuality and his most complete freedom in the political state. Man without the state is lost in the dialectical tangles of his own capricious will, but like the noble savage he discovers in this tangle the tragic contradictions which lead to his discovery of law and the common good, whereby together with others he learns to rule himself. The essence of political liberty consists in collective self-government in which private, subjective interests find their actual ends in an active, concrete spiritual order. Many critics of Hegel have accused him of conceiving the state as a static order to which the individual must submit. Hegel himself discusses such a stage in political dialectic as the apparent death of politics in the first discovery of law, a law that enslaves, but this is merely one of the tragic moments in the birth of political life from which its own dialectic helps it to recover. The judicial process by which law accommodates itself to the individual plea, the process of persuasion by which law is amended and specified, the process of rebellion by which the limits of law are discovered, all these are the vital processes of the state, and in them and only in them it has its life. The state is the order in which the individual not only finds but increases his freedoms.

There are many more aspects and moments in the philosophy of history to be found in Hegel's writings. There is the identification of history with the dialectical development of philosophical speculation; there is the establishment of history as a science in which the human spirit expresses itself in events that articulate the inner life of reason in a continuous drama; there is the identification of periods of history as moments in the development of ideas, as if societies were schools of disputation; there is the description of revolutions as if they were the inevitable tragedies of the human spirit, which gains in virtue and freedom through heroic suffering; there is the notion of history as an evolutionary process for the production of new species of men. But for the present purposes of throwing some light on the destiny of the speculative enterprise, the notion of dialectic as a scientific method ought to have special attention.

The theme of this essay has been that the various methods of science arise from the demand for unity in the sciences, a demand that has been met and partially satisfied from time to time by the working unity and continuity of method. Such methods seem to have been based on the use of two parts of science for the intensive development of the third. The ancient method falls into the pattern of subordinating facts and hypotheses to the aim of discovering and formulating principles. The method of the Renaissance seemed to be based upon the use of principles and facts for the construction of hypotheses. The modern period is marked by concentration on the discovery and accumulation of facts with the less and less recognized use of principle and hypothesis for this purpose. The parade of these methods shows imbalance and extravagance at any given time, but the whole spectacle shows a certain orderly sequence of the diverse approaches to the speculative task, a sequence that might claim continuity and a power of self-correction. The historical review even claims to reveal a method of methods. This undoubtedly is what Hegel divined in the history of science and used as the science of history.

But he also saw reciprocal relation between the dialectic of science and the dialectic of politics. Politics can be understood as the life of reason in society, and science can be seen as the concerted attempt to make reason articulate for its application in politics, as the refracting theoretical medium through which reason moves to awaken and to shed light on the body politic. But for Hegel there would be a reciprocal influence. The state would realize and appropriate science as a vital and essential part of its own life. In seeing this Hegel was not only a good historian, as may be seen in the increasing weight that scientists had given to their industrial, economic, and political responsibilities since the time of Francis Bacon; he was also prophetic in pointing to the connection between science and politics that has made both scientists and statesmen anxious for the future of freedom in both science and politics. Recently this has taken the somewhat surprising and fearsome form of state-controlled laboratories in Russia and Europe, and the climax seems to have been reached in the warlike demonstration of atomic energy by the United States government. The Russians respond to this demonstration with an announcement of the development of atomic energy for industrial purposes, and the United States answers quickly with the decision to manufacture the hydrogen bomb. If there is a dialectic of the cold war, or if cold war is a kind of dialectic, the synthesis may be the transfiguration of the earth into a new sun. It is said that Hegel went to the window as he finished *Phenomenology of Spirit* to see Napoleon ride into Berlin at the head of his troops; his comment was that he had seen the Absolute on horseback. Thus do philosophers betray the pious hope that somehow men may become wise enough to rule themselves.

At the height of the melodrama of the cold war it would seem improbable that anything wise can be said. On the other hand, the reason that at such times seems to guide history in a dream is neither asleep nor fully awake, and it will wake itself only by trying to talk. Actually, there is a voice in the cold war, distorted by passing through the so-called iron curtain, at times drowned out by the echoes of two world wars, and now jammed by the strident chorus of other voices that rush to interpret the daily news. As Hegel's dialectic traced the plot in the chaotic drama of the French Revolution, so the Marxian dialectic, which recognizes its origin in Hegel, prophetically set the plot for the contemporary world crisis. In the last hundred years this voice has embodied itself in a certain order of events which Hegel had ignored, the progressive penetration of abstract science into the processes of industrial production and consequently into those institutions of labor and commerce that surround the modern marketplace. Hegel would have seen in this process the abstractions of science coming to know themselves in institutional embodiments, the spirit of history recognizing itself in one of its creative moments. He would have agreed with Marx in seeing it as a revolution of world-historic importance, but Marx achieved a simplicity and intensity of interpretation by reducing the complexity of Hegel's dialectic to the mechanism of one of its parts. Marx described his achievement as "turning the Hegelian dialectic right side up." There are times when one wonders if history itself has not turned itself upside down, as if to anticipate the Marxian prediction.

Marx, with an eye sharpened by a study of the new science of economics, sees the dialectic of history moving between two universally recognized social classes, the rich and the poor. The alternate intervals of war and peace between these two classes are marked by the birth and death of institutions which establish temporary truces from time to time. The successive theses and antitheses of the recorded dialectic are master and slave, baron and serf, capitalist and worker. The hero of this relentless drama is labor instead of spirit, and the syntheses mark achievements of partial justice rather than freedom. The tragic irony consists in the repeated fact that the progressive achievement of justice in the succession of political states results in the reinforcement of partial slavery for the worker. As one might see in Hegel's dialectic an ideal solution in a free world government, so Marx saw in his dialectic an ideal victory of labor in a just economic system for the world. This antithesis of freedom and justice might be seen by both Hegel and Marx as the basis and demand for a new synthesis, since neither is ignorant of the validity of the opposing claims. Both of them see persistent themes in history that their followers have turned into principles of determinism, political determinism for the Hegelians and economic determinism for the Marxians, but Hegel would not be Hegelian in this respect, as Marx said he was not Marxian, when he was asked to extend his predictions.

There are many ramifications of the Marxian dialectic, the most familiar of which is the interesting proposition that the forms of historic institutions are determined by the means and methods of production, but again for present purposes it will be sufficient to see the Marxian dialectic as it proposes a method of methods for science.

Marx sees science playing a decisive historic role in the changes in the means and methods of production. As we have seen, the empirical part of science is always closely connected with the artifices of human making, therefore with the workman and the craftsman. The abstract parts of science are on the contrary connected with the liberal arts, which at least in one historic interpretation of the term *liberal,* are the prerogatives of the freeman or master. The liberal arts are also essential parts of the art of politics. This does not mean that there are not isolated cases of slaves or serfs or workmen who have been able scientists and philosophers, nor that there have not been liberal artists who were good experimental scientists, but Marx sees in the record an expropriation of knowledge, as well as the skimming off of financial profit from the worker. Perhaps the tradition of liberal education for the few who have the leisure to afford it is conclusive evidence of the expropriation.

If it is true that the combination of the liberal and useful arts in each man by rotation of functions in the monasteries is the true origin of the industrial revolution, which is marked by the application of abstract science to the productive arts, then we have in the modern period a new and more fatal expropriation of knowledge in the alliance of the university and the research foundation for the cultivation and promotion of science as against the ignorance and presumed inability of the workmen to understand their own industrial operations. The distinction between the intellectual and the laborer in the modern period is only partly moderated by a system of universal education and the existence of the engineer and the mechanic with know-how.

In terms of the historic dialectic Marx would be the last to regret this separation of function, although it seems to him to entail unjust expropriation. It is by means of the isolated and organized cultivation of abstract science that the means and methods of production have been increased in quantity and power, so that not only a just but also a plentiful distribution of the products of labor is now possible. Similarly, it is only the capitalistic expropriation from the worker of the tools of production that has built the technology which is adequate for the support of the population. Still for the dialectical historian the capitalistic system is only a stage in the historic process, a stage like previous ones, in which inherent injustice expresses itself in inner contradictions which violate the fundamental law of economics that a workman has a right to the product of his labor, in terms of knowledge as well as in terms of material wealth.

Of course Marx, like Hegel, took the themes of his dialectic from many voices that preceded him in the great conversation. His theory of the market, of money as the medium of exchange, and even the labor theory of value are taken immediately from Adam Smith's *Wealth of Nations*. This book which is often taken to be the classical origin of the modern science of economics is, as the title indicates, concerned with political economy, that is, with the role and duties of the state with respect to the production, consumption, and exchange of goods. Together with his *Theory of the Moral Sentiments* it is quite clear that *The Wealth of Nations* is concerned to find out how the operation of the free market serves the common good and how the government, properly informed, can detect and correct the operations of industry and commerce that do not contribute to the wealth and welfare of the nation. To correct a common contemporary error, it may be said that Adam Smith is advising the state how to make its necessary interference with business effective and just.

Marx sees in state power a necessary hesitancy and weakness that merely protects property and therefore the interests of the rich against the poor. But in place of state power, which fails to penetrate economic affairs far enough to see that justice is done to all citizens, Marx sees in the heart of the modern industrial process the steady growth of the almost forgotten original root of economic life, a form of the common good which was known and honored before the invention of politics. Even the original meaning of the term economics, the law of the household, memorializes it. In the family, which might grow to the size that equals the tribe, the work of the hands of many craftsmen contributed to the common wealth of the community, supplying by labor the necessary means of life, food, shelter, and clothing without the benefit of the free competitive market. The division of labor in this community was made only to articulate better the services to the common good.

Modern industry broke this tribal or feudal form and set free the social habits which had developed under it. As long as handicraft persisted, the marketplace and the institutions of monarchical government imitated the social customs of the family, but with the establishment of the factory there came a new form of human association. Workmen summoned and bought by the factory owner became members of a new community in which common work and common suffering under exploitation engendered new social sentiments and new social habits. Under the appearance of a market where labor was bought and sold in a competitive system there was increasingly recognized a new workers' commonwealth. The introduction of machinery and the methods of mass production confirmed the feelings and habits of cooperation and added a rigorous discipline of highly articulated skills: the industrial process automatically and naturally organized the poor.

As if by reflection of this socialization the owners of the factories organized themselves into joint stock companies and banks in which the common good of the rich took the place of the free market and its competitive customs. This development is often rhetorically emphasized as the intensification of the class war, as the forms of competition polarize into a single contest between capitalist and proletariat, but for Marx himself it is equally an accelerated process of socialization. The recognition of its existence and the acceptance of its apparently inexorable motion amounts to a completed social revolution. In Hegelian terms, self-consciousness with respect to the process of socialization will bring freedom, a social freedom that is equivalent to political liberty; in Marxian terms, the class warfare leads to a victory of the workers of the world and the abolition of classes.

Abstract science has not only supplied the technology and therefore the basic pattern of organization for this new community; its modern empirical development came from the engineering that went into the building of late Renaissance cities (which Galileo observed in Padua), from the mechanical clocks of the monasteries, from the mines and smelting furnaces of Europe, and finally from the combination of steam and coal that turned the shafts, pulleys, and cogs of the factories. The original inventions came from the craftsmen, but the later ones came from the laboratory, where measurement was added to experiment to supply general laws that would displace rules of thumb. Science also became socialized in this process.

When Gladstone asked Faraday what use his dynamo might have, and Faraday answered that some day Gladstone might tax it, there was a meeting of minds. The scientist knew that he was making dialectical history and the statesman knew that he might have to rule science. Inside the prophetic irony of this brief dialogue, a hundred years of dialectical history has thought and acted out its meaning. Faraday is usually understood to have intended contempt for the Oxford-trained politician, a contempt justified by a life of single-minded devotion to the search for the truth. Faraday has become a kind of saint of the freedom of science, a cult that is thus placed, ironically enough, in the Royal Institute, with Davy, the inventor of the miner's lamp, at its head. The scientist is supported in what Veblen has called his idle curiosity by government subsidy. Another later variant on the pattern is seen in the modern research foundation making astronomically large grants of money accumulated from oil for the construction of the largest reflecting telescope in the world, or the university president defending academic freedom for an institution that is endowed by large gifts from its industrial alumni or from funds appropriated by state legislatures.

Reflecting this picture of industry and government supporting and serving the ends of disinterested inquiry, there are publicly and privately supported laboratories in universities, colleges, and research institutes, industrial laboratories in the big companies, and laboratories now in all branches of government, uncovering and formulating the foundations of the future society. The role of government is taken to be that of the impartial spectator with a single vision of justice for all. Political freedom is defended from the dictatorship of the scientific planner and technocrat. Meanwhile, the labor union prides itself on being independent of politics and innocent of technology and science.

Nowhere except in the backward country of Russia have the issues of the dialectic been joined, and there the fateful leap has been made into conditions that require the doctrines developed in highly sophisticated and partly socialized industrial countries to be applied for the purpose of pioneering and developing a large area that is comparable industrially with North America in the seventeenth century. In spite of this, the result in scientific method is highly instructive. Whereas we encourage and defend the freedom of the individual scientist, or the institutes that are units of the scientific guild, to discover, test, and apply ideas and techniques and to ignore the subtle external influences that guide their dialectical meanderings, it appears that scientific method in Russia, including the choice of projects, is a matter for political deliberation and decision in the face of vital popular needs. The deliberations are said to follow an explicit dialectical pattern suggested, if not implied, by the Marxian dialectic. This allows the clear definition of the assignment, the efficient supply of equipment, the careful division and coordination of effort, and the constant critical abandonment or replacement of ideas, methods, and techniques if they appear futile or inefficient. The relevance of this dialectical materialism in science to our problems is somewhat reduced when we realize that there are no folkways or customs in Russia that support either scientific, or academic, or political freedom, all of which for two or three centuries we have fought and died for. We at least are unable to understand or believe that two or three centuries of our experience can be passed through in a generation or two by a backward people. For science we do not believe truth can be arrived at by legislation or edicts governing laboratories.

Our own dialectic, implicit though it may be, is no less puzzling. We have large associations of power collected and organized around the three themes of the dialectic, science and technology, industry and finance, and government. Each by the very pattern of its concrete activities penetrates the others, but the forced meeting of any two or all three of them, as such things have happened in two world wars, sets up a reciprocal relation of suspicion and fear which can break out on occasion—as when atomic energy was developed by a forced collaboration of all three—into widespread panic. For the first time in our history the scientists organized themselves for political action, business acted coy and retiring in the face of a new invention, and the government set up a legal monopoly for the whole enterprise. These fears and reactions are familiar to Freudian students of anxiety, but Hegel's dialectic is more illuminating than psychoanalysis. Our deeds for the time have outstripped our intelligence, and we are now involved in the dialectic by which we can come to recognize and exercise a new kind of freedom. There can be little doubt that Marx is partially right also in warning us that it is the degree of socialism at which we have arrived that we must recognize and accept.

As far as science is concerned, and it may be that it is from a critical view of science that the new light will come, we have the pragmatist's view that the validity of science rests on the working utility of the empirical methods of observation and experiment, and that we shall have light on the puzzles here only when we agree to submit more and more of our experience to these methods. We have the typical view of the mathematical physicist, like Einstein, who works to unify all physical law in one grand equation, which might then with suitable intermediate terms be applied even to social organization. And we have the view of the dialectical materialist, who would find the relevant method in science by deriving it partly from the social necessities of the time—by deciding on the socially necessary experiments—and partly by choosing among the current technological facilities available for purposes of investigation.

To the reader of Western literature it will occur, perhaps as a dream of perverse nostalgia, that the present tangle of dialectical methodology may arise from a false question. It seems that the dialectical problem arises from the question: What is the basis of validity of the empirical method? It may be remembered that the facts, the troublesome stuff of science, are only one part of its historic enterprise, and the previous subordinations of two parts to the intense pursuit of a third have each led to extravagances and paradoxes. It may further occur that this exploitation of parts may itself be a disease of the human mind, perhaps a disease of the Western mind as distinguished from the Oriental mind, which has never succumbed completely to the temptation to abstract science. Might there not be some categorical imperative of the speculative intellect which would enjoin it to see each part of science as an end in itself, and some principle of liberty, equality, and fraternity which would confer on the intellectual power of science the virtue of justice, a distributive justice to its parts and the proportional justice which would make it the medium of a single clear vision, something like the view that great poetry affords? Such a view might be the simple end for which all method reaches, but which it cannot grasp. Such a view may be achieved by one who reads science for the beauty and enjoyment it affords.

Index